Farm Lease Guide

Robert M. Dunaway
Alvin J. Morrow

ISBN 0-87069-357-3
Library of Congress Catalog No. 80-53010

Cover photograph: courtesy of the
Soil Conservation Service

Published by

Wallace-Homestead Book Company
1912 Grand Avenue
Des Moines, Iowa 50309

Dedication

To our mothers — Ellen Dunaway and Christine Morrow — land owners who rent their farms to good tenants.

Acknowledgements

A number of people have been instrumental in helping the authors get this farm lease book to you. We want to recognize their assistance.

Our technical advisors and reviewers are Robert A. Luening, University of Wisconsin Professor of Community Affairs; Franklin J. Reiss, University of Illinois Professor of Land Economics; Everett G. Stoneberg, Iowa State University Extension Economist; Robert Walters, 1980 president of the American Society of Farm Managers and Rural Appraisers and a member of the Hertz Farm Management firm at Nevada, Iowa; and Thomas E. Rittgers, a member of the Hertz Farm Management firm at Nevada, Iowa.

Several Farm Progress Publications editors also provided information and editing assistance. They are Charles Kuster, James Head, and Frank Holdmeyer, **Wallaces Farmer,** and John Vogel and Carl Eiche, **Prairie Farmer.**

We also thank Al Bull, Vice President and Editorial Director of the Farm Progress Publications for his editing assistance and encouragement.

Also, we thank Robert Anderson, a Farm Manager for Iowa Des Moines National Bank, for the information he provided.

We feel that this team approach has helped us provide you with a better farm leasing book.

Table of Contents

Introduction

As farm magazine editors, we have received more questions from landlords and tenants about farm leasing problems than about any other subject. Yet, when we have searched for answers, it seemed we could only find them in bits and pieces.

That's why we have written this book. It can help you decide if your present lease is reasonable and fair. And we hope it will show you how to design a better farm lease agreement. It should answer most—and hopefully all—your questions about renting farm land and buildings.

It is written for both landlords and tenants. That's because the best farm leases are ones that are fair to both parties involved.

Since no two leases are exactly alike, there's no easy formula. But we show a number of alternatives you can use to figure what is fair. We urge you to pencil out your own situation using several of the alternatives and your own figures.

The fairness of any lease should be judged in total rather than on individual parts of the lease. Only the parties directly involved can make the final decision on fairness. And it usually requires some give and take by both landlord and tenant to design a lease that makes both parties happy.

Our goal in this book is to give you as much help as possible. But we do want to caution you about the figures such as costs of production and custom rates. These are averages since there's no one set of figures that will apply to everyone.

Your figures may be higher or lower. So we suggest you

use your own records to be more accurate.

Also, if you use this book for several years, the figures will become outdated. So take that into consideration.

Finally, we have made every effort to make this book about farm lease arrangements as complete and accurate as possible. We hope you will find it to be another valuable tool in your farming operation whether you are the landlord or the tenant.

Robert M. Dunaway

Alvin J. Morrow

Chapter 1

Selecting A Tenant

You probably know landlords who have had the same tenant for years. A bond of mutual trust, a desire for the same farming goals, and friendship have often developed between such landlords and their tenants. In other words, they are a good match and a good team.

That's the same type of working relationship every owner would like to establish when selecting a new tenant. It's obvious that the right choice of a tenant is essential to your being satisfied with the way your farm is operated and the income it generates.

Define your farming goals

But how do you find that "right" person to be your tenant? With a big demand for the few farms available to rent each year, you'll likely have a number of choices—some good, some not so good.

The "tenant selection process" should start with an assessment of your goals and an analysis of your farm's needs.

You must have a clear understanding of what you want your role to be as landlord. For instance, do you want to be involved in management decisions or do you want the tenant to take complete charge of the farming operation?

Write out your planned role as the landlord. That will help you define what you expect from a tenant and what he might expect from you as the landlord.

Then look at the special needs of your farm. For example, you may have both cropland and some pasture. Can the prospective tenant utilize that pasture through a livestock program he presently has?

If he's strictly a grain farmer, he probably won't be able to use pasture. But he might want to rent the farm just to get the cropland. That might not be bad if you can find someone else who wants to rent the pasture. But if the pasture goes to waste, you lose potential income. It's easier to rent the whole farm to one tenant. Then you may agree to let him rent the pasture to someone else if he can't use it.

Or, as another example, your farm might be prone to soil erosion. If so, you'll want a tenant who's willing to follow conservation tillage practices.

Get help if needed

At this point, you may decide you need help selecting the right tenant. Possibly you're a widow who doesn't want to become involved in details of the farm operation. Or you might be a son or daughter who left the farm after school and haven't kept up on farming practices. There are many other reasons why an owner might not feel qualified or want to actively select a tenant. If so, you have some options.

• One of your best bets is a professional farm management firm. For a fee, the professional farm manager will choose qualified tenants and be actively involved in management of the operation. (See the chapter about professional farm managers for more details.)

• A farm consultant may also be helpful. He's an agricultural specialist in the business to help farmers solve specific problems. However, few farm consultants at this time specialize in matching tenants to specific farms. But you may want to check this possibility out.

• A qualified disinterested third party may be another possibility. This could be your county extension officer or agent, banker, production credit association manager, vocational agriculture instructor, or someone else knowledgeable in farming. Some of these people, however, will shy away from recommending or helping select a tenant because of a conflict of interest—the possibility of needing to choose between people who are their clients or whom they know quite well.

Starting the search

It should be fairly easy to locate an individual to meet the needs of your specific farm. There are many farmers wanting to expand their acreage or move to a better farm. Keep in mind that there are certain farmers better equipped with both machinery and experience to handle a certain type of farm.

You can locate potential renters by advertising in farm magazines and newspapers and by spreading your needs by word of mouth.

But you may prefer the ''low profile'' approach that many professional farm managers use in selection of a tenant. They often seek out tenants by going to businessmen in the community—elevator operators, chemical and fertilizer suppliers, and bankers—who are familiar with farmers in the area.

Local businessmen are often willing to recommend possible tenants. While the businessmen usually won't divulge personal information, they're usually fair in their appraisal of potential renters. They know which farmers are doing a good job and are financially sound.

There's another reason for taking the low profile approach to selecting a tenant. You could be deluged with interested farmers if you advertise for a tenant. In one case, more than 30 farmers showed interest in renting a single farm. To be fair to all, you should visit with each individual. And that takes a lot of time and patience.

With the demand for land to rent, you'll likely find a number of promising candidates to be your tenant. Your job then will be to sift through the candidates to pick the one best qualified for your operation. Your first "sort" may be easy. Some will immediately show themselves as top candidates while others will appear less likely to fit your standards.

Visit the tenant's farm

As a starting point, visit each promising prospective tenant's farm. Make an appointment with him and handle the visit in a businesslike but friendly manner.

Interviewing the prospective tenant and his or her family on his farm does two things. He often feels more comfortable there and more ready to discuss his position and goals. At the same time, you can make a firsthand appraisal of the person's farming ability, machine capacity, and certain other traits.

During the visit, evaluate a prospective tenant on his ability to express his views. What are his feelings about farming? What is important to him? You want a tenant who is willing to communicate with you. This visiting will help tell if your personalities are compatible.

Visit with the person's spouse, too. Her attitude can be a big factor in your decision. Many wives get involved in both labor and management decisions. And if you furnish a house, you'll want to get a feeling as to whether or not she will be happy with it.

The initial visit is also a good time to discuss farming practices. What crops does he like to grow? Is he strictly a corn and soybean farmer or does he like to rotate with some oats and forage crops, for example? How does he feel about certain farming practices such as soil conservation?

You can also look over his farming operation during the visit. How do his crops compare with others? Is he following the best cropping practices? How do his yields compare with other farms in the neighborhood? If his crops look poor but he has a lot of excuses, evaluate those excuses.

Check his facilities and equipment. Most farmers will readily show you their equipment and what they are doing on their farms. If the farmer doesn't want to show you his farm or equipment, there may be a reason. Maybe he isn't doing a good job. Maybe his equipment isn't up to date or in good repair. If he's trying to operate on a shoestring, he may not be a good prospect.

Note the tidiness of the farmstead. Is it fairly clean, picked up, and are equipment and tools in their proper places? Is the family proud of the crops and the farmstead? This helps indicate how well a man cares for his equipment and how efficient he is. Don't plan to change the person's habits once he farms for you. It probably won't work.

Can he handle your farm?

One important question you will want answered during this and/or other visits with the prospective tenant is his ability to handle your acreage and do the work on time. Discuss with him how much land he is presently farming. Also, how much help does he have available and how big a line of equipment does he have?

However, don't be overly impressed with a huge amount of equipment. A farmer may be farming too many acres in an effort to cover his investment in a large amount of equipment. In that case, he may be spread too thin to be timely on all his land.

Also, ask the tenant if he can be timely with your operation. But weigh his answer. Some tenants, intent on getting more acreage, may be overoptimistic in assessing their capabilities. What happens if there are five more rainy days than usual at planting or harvest time?

Does he farm other land? And how far away is it? Will he continue to farm some or all of it?

Closeness of farms he rents can be highly critical in your analysis. A farmer can usually do a better job if he isn't farming all over the county. If the land he farms is spread out more than 10 to 15 miles, be cautious. Even that is putting a lot of miles on the road. Of course, realize that it may be more practical for a farmer to operate two large parcels of land 10 miles apart than four small farms spread out over shorter distances.

Concern for conservation should be given a high priority. You can best determine a tenant's views by an on-farm inspection—especially if it's land he owns. On rented land, lack of good conservation practices could be at least partly the landlord's fault. Check what he's doing to conserve soil, such as terracing sloping land. If he's growing soybeans on a hillside without suitable conservation practices, that's a red flag that he's not concerned about saving the soil. Also ask him about the present conservation practices on the land he farms.

References may be important

During the selection process you'll also want to get someone else's opinion on how good a farmer and renter a prospective tenant is.

Checking references is a good way to find out. Visit with his past and present landlords.

But tell the tenant you're going to do that. Ask him for names of his landlords and whether he objects to your visiting with them. You may also want him to inform those people that you will be calling them.

Every party to a new arrangement should be aware of what's going on. It's good business on the part of the tenant to let his other landlords know he plans to take on other land.

This also does another thing. It alerts the other landlord that someone else wants this man to farm on his land. The landlord who is presently working with him may say, ''I don't think he can really handle another farm.''

However, there are some landlords who want the tenant to farm only their land. That's pretty restrictive. If a man can handle more land, there's no reason to have that type of attitude.

The other landlord can also give you some insight into a prospective tenant's farming ability, integrity, and ease to work with.

Professional farm managers require a financial statement from prospective tenants. That includes their accounts and debt load.

Asking for this type of information can become a ticklish situation for a landlord who's a member of the same community. A tenant may not be as free with financial information with a neighbor as he would be with an outside source. Nevertheless, you should be concerned about the financial status of a potential tenant. You don't want a tenant who will be undertaking more than he might be prepared to handle. So don't be afraid to ask for a financial statement. His banker may also be willing to say the "tenant is sound" or "he's a little shaky" without going into detail.

Finally, if you have any reason to question the person's honesty, look elsewhere for a tenant.

Chapter 2

Selecting A Landlord

Selecting a landlord in most cases does not need to be a difficult process. It boils down to three main steps:

(1) Assure yourself that the prospective land or unit you want to rent fits your needs.

(2) Check out the landlord. He should be the type of person you can work with for the benefit of each.

(3) Sell your farming expertise to the landlord to convince him that you're the best person to farm his land.

Does the farm fit your needs?

The starting point in the landlord selection process is obviously the farm he has available to rent. Determine if it suits your needs. The prospective farm should blend in and be a viable part of your existing operation.

Obviously, if you're a grain farmer, that's the type of farm or land you want to rent. A farm with lots of pasture, for example, does little good if you're not already in the cattle or sheep business. So ask yourself: Do I really need this added land or do I just want an extra 80 or 160 acres to add to my list of farms?

Closely related to your need for the farm is the profitability of the farm. You should be able to make money off it or it shouldn't be considered. The land's yield potential, along with how much more effectively you can allocate the use of equipment and labor, should let you pinpoint closely the profit potential.

If you don't own land or won't be renting additional land, be sure the rental unit has adequate income potential. Some farms have limited potential. That may be why the farm is for rent.

Also look at the location of the farm in relation to your other farm or farms. You want the farm close enough so you can be timely with field work. Generally, you'd be spending too much time on the road if your farm is 15 miles or farther from your base operation or other farms you operate. But keep in mind that it is probably more practical to operate two large parcels of land 10 to 15 miles apart than four or five farms spread out over a shorter distance.

There are other factors you may also want to consider. Does the farm have buildings that you need? Are the fields large enough to accomodate your equipment?

Checking the landlord

Once you've determined that you can use a prospective farm, then check out its owner. He should be someone you can work with satisfactorily. That is necessary for a successful leasing arrangement. You can usually measure the acceptability of a landlord by the following criteria:

• Reasonable. This trait tells you how cooperative your landlord is and how willing and easy he will be to work with. It's also a sign of whether the landlord will give and take to make the lease fair to both parties.

• Communications. The landlord should be willing to visit with you about his farming goals. From the beginning, you want to have an open expression of views with the landlord. And you want to do this often. Don't be afraid to ask a lot of questions. However, don't force him or her to defend the present operation or you may be considered ''pushy.'' If the landlord is sincere, he will appreciate this type of questioning.

The visit with the landlord will be valuable to your assessment of him. You might, for instance, not consider him a good prospect if he's interested only in squeezing every dollar out of the operation he can. Will he go along with a build up program? Or would he expect you to do that by yourself?

• Understand farming. You avoid some potential problems if your landlord has a good knowledge of farming. You can tell quickly when visiting with him or her. One tenant, for example, tells of a landlady who brought out a quart of herbicide and asked him to spray all the bean acreage with it.

There are also times when the lack of knowledge makes it difficult for a farmer to relate with some businessmen who do not have a farm background. For instance, a businessman might not realize the need to spend money on soil conservation practices or tiling. Thus, it's important that you determine this in the interview.

Sometimes you're going to need to rent from someone that isn't knowledgeable about farming. In that case, it's best to establish from the beginning what practices may need to be done. Although the person may not know farming, if they're reasonable and they have good judgment, they will understand and go along with improvements such as soil conservation practices.

We consider three traits—being reasonable, willing to communicate, and understanding farming—as the most important for a landlord to possess. In addition, you might want to consider others such as honesty, sufficient capital or credit to make farm improvements, good judgment, openmindedness to new practices, and respect for a tenant's privacy.

Selling your farming expertise

Once you've picked a farm and a landlord you may still need to convince the landlord that you're the best person available to be his tenant. In other words, you may need to sell yourself as a farmer.

This has become necessary with the competition for rental land in many areas of the United States. For example, we know of one situation in central Iowa where 50 farmers were interested in leasing one farm that became available. While that large amount of interest may not be typical, nevertheless tenants are finding fewer opportunities to rent land and more competition for land that is available.

The best advertisement you can probably give for renting another farm is the job you're already doing on your own farm. Your prospective landlord will check you out one way or another, probably by visiting your operation. He may not be interested in you if he sees weeds in your corn and soybean fields, fence rows that aren't clean, or an untidy farmstead and yard.

A prospective landlord may be impressed by an impressive cash flow budget. Show him on paper how well you've done in the past with your farming operation and what your cash flow plans are for your operation for the future and for his farm. A landlord will not want someone who is spending more than he expects to make or who is living off inflation and depreciation.

A landlord likes a tenant who understands technology. Someone who's up on modern farming practices. Someone who knows what herbicides to use, for example.

The amount of equipment can also be an important selling point. So may be the condition of your equipment. The landlord will probably be more impressed by your background to get the job done. They feel equipment is really secondary to being able to handle finances and your understanding of farming. It's what you can do for him or his land with your equipment that really counts.

The involvement of your wife can also turn the tables in your favor. Usually, the landlord likes to see a tenant's wife who is enthusiastic and knowledgeable about what her husband is doing.

Summary

You'll have a firm foundation to build a lasting landlord-tenant relationship once you've gone through this process. You'll feel good about your landlord and he'll feel good about you. You'll understand each other or at least what each others goals are.

Chapter 3

Use A Written Lease

"I've known my tenant (or landlord) for years. I don't want him to think I don't trust him by asking for a written lease agreement."

That's a common dilemma. But putting your lease agreement in writing should be considered a sound business practice, not a lack of trust or confidence by either party. And, in fact, both landlord and tenant should want a written lease. It may even be that both want a written lease, but that neither has found the nerve to suggest it.

With a first time lease arrangement, it's easier to stipulate that you require a written lease as a sound business practice. Simply point out that it will be an advantage for both parties.

Let's take a look at some of the reasons you should use a written lease. There may be some things you've never thought about before that provide protection for both landlord and tenant.

• A written lease serves as a record that you can refer to if either landlord or tenant is uncertain about what they agreed to on a specific detail. That can help prevent disputes if their "recall" isn't the same.

People tend to have a "selective recall syndrome." In other words, they sometimes remember only those things that are to their advantage. That doesn't mean one person is trying to take advantage of the other. It simply means it's easy to forget exact details six months to a year after the agreement was made. That's especially true in lease arrangements where there are many things discussed in bargaining.

• A written lease protects the heirs if either party should die.

If your landlord—or tenant—dies while the lease is in effect, imagine trying to deal with his relatives without a written lease. They may be strangers to you. And they may not have any knowledge of farming—but may think they do. You could end up in a real hassle that could be expensive.

• A written lease can spell out how to make adjustments when conditions arise that makes a change desirable. For example, suppose the tenant is to pay for application costs on chemical treatments. But because of wet conditions, he isn't able to make an applicaton. Aerial spraying might be desirable, but the tenant can't afford to stand the whole cost. The written lease could provide for a way to make an adjustment.

• A written lease provides both landlord and tenant a better guarantee that they will carefully consider all parts of the lease before they accept it. It may sound good when you talk about it, but how will it look on paper? Having everything in writing and then studying it before you sign may help keep you out of a lease that might be hard to live with.

• When details of the farm operation are spelled out in the lease, that document serves as a partial history of the farming operation.

You can refer to that lease in later years to see what was done and how it was done. It may remind you of things that worked well—and some that didn't.

• The written lease can specifically spell out the period of the lease arrangement. Some states don't allow an oral agreement to continue beyond one year. But in the written lease, you can spell out that the lease is to continue for two or more years if you want.

Any one of these arguments for a written lease can easily justify the time and effort it takes. Some sample lease forms are reprinted in the back of this book. Your state university, county extension office, or your attorney can also probably supply you with lease forms. Don't let that stop you from having a written lease if one party refuses to use an attorney.

Objectives of a lease

When you write a farm lease agreement, there are basically four major objectives you'll want to consider.

- The lease should arrange for a fair division of the income and expenses between the landlord and tenant over the long run.
- The lease should be aimed at making the farming operation profitable for both the landlord and tenant.
- The lease should give as much assurance as possible to a good tenant that his lease will be continued through a period of years if he does a good job.
- The lease should give the landlord assurance that his property will be well cared for. Fertility and conservation practices should be spelled out, for example. Maintenance, repairs, and weed and insect control are also important details in most leases.

If the lease isn't fair to both landlord and tenant, the one on the short end will often try to make it up by taking shortcuts. For example, the tenant might cut back on fertilizer rates and drain the fertility. The landlord might not be willing to spend money for land improvements such as tile, for example.

Here's a more detailed checklist of questions that can help you spot potential problem areas in a farm lease:

(1) Does the lease encourage the most profitable long term operation of the farm?

(2) Does it encourage the most productive use of capital by both parties involved?

(3) Does the lease encourage the tenant and the landlord to run the farm most efficiently?

(4) Are the returns to the landlord and tenant in equal proportion to their contributions?

(5) Are the best available farming methods being used and encouraged by both parties?

(6) Is the lease written so that soil productivity will be main-

tained and useful long term improvements are made?

(7) Does the lease contain provisions whereby new and needed improvements to fences and facilites will be made?

(8) Does the lease give enough details and provide for special problems that may arise?

(9) Does the lease clearly state how income and expenses will be shared?

(10) Does it give legal protection to both parties involved?

Besides these, a complete lease should also contain a time for review; a termination date; details on how, when, and where the landlord's share of the crop or cash should be delivered; what production items are to be shared; which crops will be grown and their rotation; how crop residue will be handled; conservation methods to be followed; and responsibility for weed control in fence lines and around buildings.

An arbitration clause can be helpful in case the landlord and tenant can't settle their differences themselves. However, a carefully written lease will greatly reduce the chance that arbitration will be needed.

A four or five page lease can't cover all the possible conditions that might develop. But it does set a framework that can be helpful in resolving differences.

The written agreement should be clearly understood by all parties involved—both husbands and wives—before it is signed.

Rental terms need to be revised periodically to keep them up to date. Arrangements for this should be made in the written lease. Changes in the cost of inputs, plus changes in technology, may sway the fair shares one way or the other. So be ready to modify the rental terms to keep the agreement fair to both parties. Review the lease every year if it is a one year lease.

Finally, the small cost in terms of time and money that it takes to make a written lease will usually pay off handsomely in terms of fewer uncertainties, fewer disputes, and better farming practices that increase the profits.

Chapter 4

Bargaining

Tradition—what has been done in the past—may be a good starting point for developing a fair farm lease agreement. But no two farms, no two landlords, and no two tenants are exactly alike. Also, rapid economic and technological changes may make tradition a poor way to determine what is fair. So you will have to look at your own specific situation closely to make sure yours is a fair lease.

Shoot for a lease that is fair overall rather than getting hung up on individual details. If one party has a special concern—say on soil conservation practices—give a little on that point in return for some advantage to you elsewhere in the lease arrangement.

Quality of the land, the farm economy, and the management ability of the tenant are all big factors in how the final lease terms should read.

Bargaining then becomes the key as you look at these factors. This involves the landlord and tenant discussing each provision of the lease, describing what each feels is fair, and then compromising, if necessary, to reach a mutual agreement on each provision. The primary goal for both landlord and tenant should be to be fair while making a good or reasonable profit.

Land productivity

There's obviously a lot of difference in land in terms of how it produces. For example, some land has the potential to average 100 bushels of corn per acre. More productive soil may average 135 to 150 bushels an acre. But here's the kicker: It doesn't cost much more in terms of labor, machinery, and fuel to grow 150 bushels an acre than to grow 100 bushels. Fertilizer, chemical, and seed costs may be only slightly higher.

Typically, cash rent has been adjusted to take care of that difference. But 50-50 crop share leases usually haven't.

For example, the chapter on crop share leases indicates that variable costs are only about $8 more for the landlord and $11 more for the tenant to produce 130 bushels of corn as compared to 100 bushels. But each one gets an additional 15 bushels of corn.

The big question is, ''Who is getting the best deal?'' If the 50-50 lease arrangement is fair to the tenant on the average producing land, the tenant farming the higher producing land is probably getting a better than average deal.

But it could go the other way. The tenant farming the low producing land may be having a hard time making a profit with the 50-50 arrangement. There's a tendency for some landlords to rate their land somewhat higher than an unbiased person would.

It all boils down to the fact that there may be a need for more negotiation. Both landlord and tenant need to look at their actual input costs. The tenant needs to add up his costs for machinery, labor, fertilizer, seed, and chemicals. The landlord needs to do the same thing, looking at his variable costs and a fair return for his fixed investment.

Then, the negotiations might aim for a return more in proportion to inputs provided rather than simply the 50-50 share. It might be that the tenant on the lower producing land should get a little more than 50% of the crop for the lease to be equitable. On the high producing land, it might be the landlord who shoots for a little bigger share of the crop. Of course, there may be other ways to make adjustments in order to keep the 50-50 arrangement.

Another alternative with that high producing land is for the tenant to shoulder a little more of the input costs. For example, he might agree to pay all application costs in the future if he hasn't been doing so in the past. Or, he might pay all the harvest cost. On low producing land, it might swing the other way.

Many landlords might be satisfied with the present arrangement if the tenant would agree to do some additional weed control in fence rows or around buildings, for example. That's part of the bargaining process.

Or in some cases on high producing land, the tenant can pay a small cash rent—say $10 per acre—to keep contributions in line with a 50-50 lease.

These are some of the things that can be negotiated if either the landlord or tenant feels he's getting the short end of the deal.

The farm economy

Production costs have continued to climb for both landlord and tenant in crop share lease arrangements. Landlords are contributing a more valuable asset as land values have continued to climb. Those increases have tended to change some traditional leasing patterns.

One changing pattern is that tenants sometimes take care of all harvesting costs rather than the landlord paying half as was often the case in the past. This helps compensate the landlord for providing more valuable land with increasing land values.

But on the other hand, input costs such as machinery, labor, fertilizer, and chemicals have also greatly increased the tenant's costs. Therefore, he may feel that his added costs have kept pace with or even exceeded the landlord's increased contributions.

Again, a thorough study of the inputs by both landlord and tenant is a valuable first step to see if additional negotiations are necessary.

Other factors

Modern technology has brought many changes to farming. For example, broadcast application of herbicides has reduced the amount of cultivation, a cost normally assumed by the tenant. Since the tenant saves time and machinery cost, the landlord might agree to share the chemical costs but leave the application cost up to the tenant. Or, the landlord might agree to pay only half the cost of chemical for a band application and ask the tenant to pick up the additional cost involved in a broadcast application.

Another example: Where a landlord agrees to tile wet soils, the tenant might agree to give the landlord a little additional cash rent or a little bigger share of the crop. Or, he might agree to do some additional fence repair or weed control chores.

The bargaining position of the parties involved may shift the final lease terms away from tradition, too. A tenant may be willing to negotiate more in the landlord's favor to get a farm that he really wants. A landlord, on the other hand, may do the same thing to get a certain tenant he especially wants.

Risk factors may still be another area for bargaining. For instance, if a farm is prone to flooding, the landlord may be willing to concede more to a tenant.

Bargaining solutions

Once you've compared the input contributions by both parties, you may find that your traditional lease arrangement is out of balance. Then, you have two possible solutions:

(1) Income can be shared in direct proportion to the inputs contributed by each party. For example, if the tenant contributes 60% of the inputs, he would receive 60% of the income.

(2) You may want to adjust inputs and continue with the traditional percentage shares to each party. For example, if the landlord is now providing 55% of the inputs with a 50-50 lease ar-

rangement, the tenant might pick up some additional costs that the landlord now pays so that it becomes a 50-50 arrangement on both inputs and income.

Summary

Even two farms that appear to be identical probably aren't when it comes to a lease arrangement. Each situation has to be studied carefully to develop a lease that gives each party a fair return for his/her contributions. So don't be afraid to throw tradition out the window if you have to in order to develop a lease that's really fair.

Chapter 5

Legal Aspects Of Farm Leases

There's an old adage, "an ounce of prevention is worth a pound of cure." That's why we've included this chapter on the legal aspects of farm leases.

The best way to avoid possible legal snarls with farm leases is to recognize potential problems, then correct them while developing the lease agreement.

Normally the need and desire to develop and produce a fair lease overshadows possible legal technicalities. That's probably the way it should be. Nevertheless, tenant and landlord alike should have a list of legal factors to check before completing the lease agreement.

The accompanying checklist should prove helpful for that purpose. Most major legal areas to be concerned with are given. However, we have not gone into specifics since laws in each state may be different. Also, it would take volumes of books to cover all the legal ramifications of farm leases.

You'll want to check with your lawyer to tailor these legal concerns to your particular situation.

• Legal essentials of a written lease. To be legal, the written lease must meet five criteria.

(1) It must name a specific landlord and a specific tenant.

(2) It must be signed by both parties.

(3) It must specify a definite period during which the farm is to be leased.

(4) It must contain an adequate description of the property.

(5) It must provide for payment of rent, and how the rent is to be determined.

• Oral agreements. The law in some states says that an oral agreement is as binding as a written lease. However, the oral agreement does not put the lease arrangement on a business basis and leaves room for many potential legal problems.

In some other states, however, oral leases of farmland are generally not binding because of the Statute of Frauds. However, once the tenant is in possession of a farm under an oral agreement, he either by statute or common law is given the status of a tenant from year to year.

Though state statutes vary, virtually all provide a notice period for terminating a tenancy from year to year. This means that if notice is not given to the tenant by that date he becomes a tenant for another year—hence the phrase "year to year tenancy."

• One year leases. Many leases are of one year duration. Most states have laws that cover the termination of leases or their automatic renewal, depending upon the lease.

If a written lease has a fixed time of termination, it shall cease at the time agreed upon.

Provisions for the automatic continuation of a lease from year to year may also be written into a lease.

Generally, if a tenant with a one year lease does not have it renewed in writing or it does not contain a notice provision, depending on state law, the tenant becomes a tenant from year to year. As such, he becomes the tenant for the following year and is subject to termination notice for that type of arrangement.

Be sure to understand how lease termination works in your state. It's one of the most important legal aspects of a farm lease.

• Long term lease. Some states stipulate that the duration of a lease cannot be longer than a certain number of years. The

Iowa Constitution, for example, has set a limit of 20 years on the length of a lease agreement.

Also, long term leases may need to be recorded with your county recorder. Iowa statutes, again, require the recording of leases that have a term of 5 years or longer.

• Termination of life estates. In some cases, a farmer wills his/her farm as a life estate to someone to use as long as he/she lives. This person can lease the farm and receive the income from it until his/her death.

Some states now have laws that prevent a farm lease from being terminated upon the death of the person willed the life estate. This protects the tenant from having a farm taken away in the middle of a crop season when a person with a life estate dies. The lease at the time of death remains in effect until the current lease ends.

• Tenant's rights to remove fixtures. Often a tenant may add facilities to a leased farm, say a grain bin. The question then arises—can the tenant remove the facility when the lease is terminated.

While the landlord and tenant should agree beforehand on what improvements can be removed on termination of a lease, most states have laws which also set guidelines.

The rule generally followed is that a fixture is removable when the parties intend it to be and when it can be removed without material injury to the land or other buildings.

• Subleasing. In some situations a tenant may want to lease or assign his farm to someone else. Landlords should be concerned with this practice because someone might end up operating the farm who isn't as good as his tenant. For that reason, many written leases have a provision against subleasing or assignment without the written consent of the landlord.

States vary in at least three ways on the legality of subleasing. Most states do not have laws that cover subleasing. Thus, if there are no provisions in the lease, a tenant can in some states sublease.

In Illinois, for example, the tenant can sublease or assign a rented farm to someone else unless there is a written lease that reserves the right to the landlord.

In other states, common law says a leased farm cannot be subleased or assigned without the landlord's consent.

Still other states have statutes that prohibit subleasing without the landlord's consent.

• Rights of tenant to reimbursement for expenses. Most states do not have laws that provide for reimbursement of the tenant for unexhausted improvements. This should be covered in the lease. If not specified in the written lease, the tenant has no right to recover expenses if he doesn't get the consent of the landlord or if the lease doesn't give him the right to make the improvement.

There is also the question of reimbursing the tenant for the carryover value of fertilizer and lime when a lease is terminated. You'll find some good guidelines to follow for such situations in the chapter on crop share rent.

• Can a tenant recover losses from livestock infected with a disease already on the leased farm? The law in most states says generally not, unless the tenant can prove that the landlord knew of the infection, and failed to warn him of it. So without knowledge that the landlord knew a disease existed or without a written lease which states the premises is free of disease, a tenant likely cannot recover losses from animals that become diseased due to an infection on the leased farm.

• Rights of the tenant to use sand and gravel on a leased farm. A tenant can usually use sand and gravel located on the premises only for improvements on the farm but not otherwise.

But before taking sand and gravel for improvements, the tenant should check with the landlord. He, for example, may have sold it to a sand and gravel company, and thus neither the owner or tenant has a right to it.

The written lease may cover use of sand and gravel found on the leased farm, but seldom does. Nor do leases usually provide for the use of trees on a farm. If lumber or logs from a marketable tree are needed for some purpose on the farm, the tenant could probably make use of them. But the tenant couldn't use the marketable trees for his own purpose or to sell. There may be provisions in the lease that restrict the use of trees by the tenant.

• Tenant negligence. When a tenant takes possession of a farm he has complete control of the farm subject to any reservations the landlord might have made in a written lease.

There may be certain common law rules which give the landlord the right to come on the property to collect rent, inspect and repair the real estate, and show to a prospective buyer.

The tenant is the one who decides whether hunters and fishermen can come on the farm. He is the one who sues for trespass. If an invitee or licensee is injured on the farm because of the tenant's negligence or the condition of the premises over which the tenant has control, then the tenant, not the landlord, is liable.

However, there are situations under which the landlord can be held liable. For instance, injury may result from a condition on the farm which obviously may harm a person rightfully on the land, but which the tenant has no opportunity to correct. Or the condition may be something which a tenant would not be able to correct. The answer is that both parties should maintain liability insurance, the most important insured being the tenant.

• Summary. There are legal rules under which farm leases must operate. Check with your attorney to determine which rules apply in your state.

Chapter 6

Crop Share Rent

Crop share is the most common way farmland is rented. Stripped to the basics, the landlord and tenant each provide part of the inputs to produce the crop. Then they each get a share of that crop when it's harvested.

There's less risk with crop share for the tenant as compared to cash rent. He can still lose his share of the expenses plus his time in the event of a complete disaster. With cash rent, he risks more because he pays all the expenses plus the cash rent he agreed to even if there's no crop to harvest.

Since the landlord shares the risk with crop share, he usually makes more money in the long run than he would with a cash lease if the tenant is a good farmer. That's because he gets to share the ''extra'' in especially good years with a crop share lease. Because he shoulders more risk, a tenant usually won't agree to as much cash rent as the landlord might expect to get from crop share in an average year.

The beauty of a well designed crop share arrangement is that, over the years, profits average out fairly well for both landlord and tenant. If it's profitable for one party, it's profitable for the other. Plus, no one is usually wiped out in a bad year.

Since each person provides different inputs in a crop share lease it's smart to figure closely who should pay for what. As land values and production costs change, so do some of the common crop share guidelines. It makes sense to review the lease terms regularly—probably every year.

How to split costs/returns

The most common crop share lease for corn and soybeans is 50-50, where the landlord and tenant each get half the crop. The split may be different on other crops. On lower quality land, the landlord may have to take a smaller percentage of the return since production costs remain relatively the same as for better land while production is less.

Most production costs other than machinery and labor are normally shared the same way the crop is shared.

In a typical 50-50 arrangement, the tenant furnishes labor, most of the management, machinery and equipment, and fuel. In turn the landlord provides the land and sometimes some facilities and pays property taxes, insurance, upkeep, and repairs on his inputs.

Let's look at how other costs are normally shared in a 50-50 crop share lease.

Seed cost is almost always an even split between a landlord and tenant.

Cost of fertilizer is also equally split. But the cost of application may vary. In most cases, the landlord pays half the cost if it's custom applied. But if the tenant applies it, the landlord usually pays none of the application cost. Be sure this is spelled out in the lease agreement so there is no misunderstanding.

The reasoning is that both the landlord and tenant gain from custom application if it helps get the crop planted earlier. Also, it may not be feasible for the tenant to own specialized fertilizer application equipment. This may be a bargaining area for both parties.

Carryover value of fertilizer is considered in some leases—to reimburse the tenant when a lease is terminated. This could be especially important if the tenant comes to a farm low in fertility and builds it up.

A fair arrangement might be for the landlord to reimburse the tenant for 30% to 40% of the P and K he paid for the previous year if the fertility level was low when he came. Most of the N was used or will be lost by the next season. Another possibility is for the landlord to pay more of the cost to build up the fertility level. Then there wouldn't be any reimbursement if the tenant leaves.

If there was a good ongoing fertility program when the tenant started renting the land, he reaped benefits from it. Therefore, he's probably just leaving the land in as good shape—fertility wise—as when he started farming it. Therefore, he shouldn't expect any reimbursement for the carryover fertilizer.

Lime costs were once assumed to be a landlord expense. But since most people now consider lime as they would other fertilizers, the tenant now often pays half the cost in a 50-50 lease, especially if it is a longer term lease.

Since lime has a longer term carryover value, a tenant who shares in a liming expense should aim for reimbursement if he leaves that farm a year or two after application.

Lime normally has about a 4 or 5 year life. Of course, this will depend on the amount applied in relation to the need as shown by a soil test. Here's a table to show how much of its value would still be left at the end of each year on a 5 year life.

Years after Application	% of value remaining at end of year
1	80%
2	60
3	40
4	20
5	0

For example, suppose a tenant pays $3000 for lime knowing it will take 5 years for him to recoup his investment. If he farms that land for only 2 years, the landlord may agree to pay him for the ''unused'' part of his original cost, 60% in this example, or $1800. This should be spelled out in the lease.

Trends on paying herbicide costs have also changed. The tenant and landlord in a 50-50 share lease have normally shared the herbicide cost for a row or band application. If the tenant wanted to use a broadcast or blanket application to trim cultivation time, he paid the extra cost.

But with narrow rows, the band application covers most of the field anyway. So many landlords now share the cost of the chemical equally since a total weed control program increases yields too. That's especially true if rain keeps the tenant from cultivating or if the field or farm is especially weedy.

The landlord and tenant should also consider the fact that the tenant will have definite savings here. The broadcast application generally cuts his cultivating time and expense. If the landlord pays half the cost, the tenant might agree to use the time saved to do a better job of fence row, roadside, and farmstead weed control, for example. This is another place you can do some bargaining—some give and take.

One situation where the landowner should pay at least half the herbicide cost is when the farm is weedy when the tenant starts to farm it. In fact, if herbicide cost is especially high, the landlord might have to agree to pay more than half the cost until the weed problem is under control.

Expect the unexpected and cover yourself in the lease agreement.

Suppose, for example, a rainy spring leaves fields too wet for ground rigs and weeds get out of control, or insects become especially bad late in the season.

Maybe the tenant has agreed to take care of herbicide and insecticide application costs. He may not be able to justify an aerial application since he'll only get half the benefit.

But, if the chemical isn't applied, both landlord and tenant will suffer. This is a case where the landlord may go ahead and agree to pay half the application cost.

At harvest, the tenant stands most of the costs.

The trend in soybeans is toward the tenant taking care of the full cost of harvest whether he uses his own machine or hires the work done.

For shelled corn harvest, the landlord still sometimes pays the tenant an amount equal to the custom shelling rate on his share of the crop. Or, if it's custom harvested, the landlord may pay half the harvest costs. However, the trend is more toward the tenant assuming all the corn harvest cost.

Hauling grain to storage is normally the tenant's job. However, for an especially long haul, there's a good argument for the landlord's paying part of the cost.

The rental agreement should specify a delivery point. Usually, that's the closest market or elevator. So, if the price is better someplace 10 miles away and the landlord wants it delivered there, he should pay the tenant for the extra time and mileage or hire someone else to make the haul. In some communities, the landlord pays all his own delivery cost.

Storage facilities on the farm are often shared equally by the landlord and tenant. The tenant provides the labor and management for putting the crop in storage. Later costs to take the landlord's crop from storage to market is the landlord's expense.

In regard to grain drying, if the dryer is part of the storage unit on the farm being rented, the landlord furnishes the dryer. The tenant may pay the landlord for use of the dryer. If a portable drying unit is used, it may be jointly owned. Or, either party may own it and charge the other for using it to dry his share of the crop. Fuel and power costs for drying are normally shared the same as the crop is shared. The tenant provides the labor for drying if the landlord provides the dryer.

The specifications for harvesting, drying, storage, and delivery of grain to market are the areas of greatest bargaining between landlord and tenant when developing a fair lease. They offer a lot of potential for give, take, and tradeoffs in lease agreements.

Government program payments

Quite often, crop share leases overlook some other important provisions.

For instance, how are government program payments split? Uncle Sam specifies that such payments are to be shared on the same basis as crops from the land being rented are shared. So in a 50-50 crop share arrangement the tenant and landlord each get half.

If it's in a set aside program, the tenant normally takes care of tillage, seeding, and weed control on the set aside acres. Costs for fertilizer and seed are shared the same as the payments are shared.

If the tenant has livestock, he may want to pasture the cover crop on set aside acres during the allowed time. Then he will usually furnish the seed, fertilizer, and chemicals needed for that crop. Be sure to agree on these things in advance.

Crop insurance can be carried by either party on his share of the crop. You don't really need to put it into your formal agreement. But make sure each one understands that he is responsible for it on his share.

Other splits to consider

There's room to maneuver with crop share leases. Shoot for a lease that's fair to both tenant and landlord. Give and take bargaining makes a lot of sense.

You may find a good argument for the split on both income and expenses to be 3/5 for the tenant and 2/5 for the landlord. Other possible splits are 2/3—1/3, and 3/4—1/4.

Look at land value, cost of inputs, and expected yields to determine which split is best for you.

Here are some estimates that will help for corn and soybean production. Just remember that these figures are averages and that they will likely become out of date. We suggest you use them as guidelines only. Your own actual cost figures will be much more accurate for your own specific land.

Variable cost estimates of corn per acre

Yields	80 bu.	90 bu.	100 bu.	110 bu.	120 bu.	130 bu.
Seed	$10	$11	$11	$12	$12	$12
Fertilizer & lime	24	26	29	32	36	40
Herbicide & insecticide	13	14	15	15	15	16
Machinery var. costs	17	17	18	19	20	21
Crop drying, ins., misc. & int.	18	19	20	21	22	23
Total Variable Costs	$82	$87	$93	$99	$105	$112

Gross returns per acre from corn

Yields bu.	Corn price $2.00	$2.25	$2.50	$2.75	$3.00
80	$160	$180	$200	$220	$240
90	180	203	225	248	270
100	200	225	250	275	300
110	220	248	275	303	330
120	240	270	300	330	360
130	260	292	325	357	390

Using a 110 bushel corn yield and $2.50 corn price, here's

how income and expenses might be split under the various leases.

| | 50/50 | | 3/5-2/5 | | 2/3-1/3 | | 3/4-1/4 | |
	Tenant	Landlord	Tenant	Landlord	Tenant	Landlord	Tenant	Landlord
Income	$137.50	$137.50	$165.00	$110.00	$183.33	$91.67	$206.25	$68.75
Expenses								
Seed	6.00	6.00	7.20	4.80	8.00	4.00	9.00	3.00
Fertilizer & Lime	16.00	16.00	19.20	12.80	21.33	10.67	24.00	8.00
Chemicals	7.50	7.50	9.00	6.00	10.00	5.00	11.25	3.75
Machinery-Variable	19.00		19.00		19.00		19.00	
Crop drying, etc.	10.50	10.50	12.60	8.40	14.00	7.00	15.75	5.25
Total costs	$59.00	$40.00	$67.00	$32.00	$72.33	$26.67	$79.00	$20.00
Return	$78.50	$97.50	$98.00	$78.00	$111.00	$65.00	$127.25	$48.75

For the tenant, the return figure has to cover his fixed or ownership costs for machinery, labor, and management. Any income above those costs would be profit. For the landlord, this is the return for his land investment and the income to pay for land costs such as taxes, insurance, and upkeep.

You can see that as the tenant takes on more risk—a bigger share of the costs—he stands to make more profit in a good year with the other than 50-50 splits. But he also stands to lose more in a year when yields or price are lower. Often, if the tenant gets two-thirds or more of the crop, he assumes all expenses other than land costs.

The landlord may not like or accept a smaller share of the crop—especially on land that can be expected to produce 110 bushels of corn.

A bigger share for the tenant and less for the landlord is more likely to be fair on lower producing land. Here is a comparison of a 50-50 and 2/3—1/3 split on lower value land—land expected to produce 80 bushel per acre corn.

	50-50 Tenant-Landlord		2/3-1/3 Tenant-Landlord	
Income	$100.00	$100.00	$133.33	$66.67
Expenses				
Seed	5.00	5.00	6.66	3.34
Fertilizer & lime	12.00	12.00	16.00	8.00
Chemicals	6.50	6.50	8.66	4.34
Machinery-Variable	17.00		17.00	
Crop drying, etc.	9.00	9.00	12.00	6.00
Total cost	$49.50	$32.50	$60.32	$21.68
Return	$50.50	$67.50	$73.01	$44.99

With the tenant's investment in machinery plus his labor input, the $50.50 return on a 50-50 lease may not leave much profit. He does have the advantage of lowering his machinery cost per acre by spreading his machinery investment over more acres by farming the extra land. But if he has a choice, the tenant will be better off to take on a higher producing farm.

The landlord may face having a less desirable tenant with the 50-50 arrangement since the tenant's return is low. That could lead to an even smaller yield.

He may prefer to give a better farmer—tenant—a better deal such as the 2/3—1/3 share lease. Of course, he has to consider his costs for taxes, insurance, and upkeep of the farm, plus, he needs a fair return on his investment. That's where bargaining might come in.

For example, the landlord would prefer a 50-50 split. The tenant might say he needs a 2/3—1/3 lease. And they might settle on a 3/5—2/5 agreement.

Supply and demand on land is a big factor. Typically, there has been a strong demand for farmland that's available to rent. So landlords should look beyond the dollar figures on paper. Quality of the tenant may be more important—and more profitable in the long run.

Here are cost estimates for soybeans. You can use these figures to estimate landlord and tenant returns for different share arrangements. Again, they are averages. Your own actual figures will help you be more accurate.

Variable cost estimates of soybeans per acre

Yield	25 bu.	30 bu.	35 bu.	40 bu.	45 bu.
Seed	$ 9	$10	$10	$11	$11
Fertilzer & lime	11	12	13	14	15
Chemicals	11	12	13	14	15
Machinery var. costs	10	10	11	11	12
Misc. int., crop ins., etc.	10	10	10	10	10
Total Variable Costs	$51	$54	$57	$60	$63

These are approximation of estimated costs since no exact estimations are available by yield breakdown.

Gross returns per acre from soybeans

Yield bu.	Bean price	$4.50	$5.00	$5.50	$6.00	$6.50	$7.00
25		$112	$125	$137	$150	$162	$175
30		135	150	165	180	195	210
35		157	175	192	210	227	245
40		180	200	220	240	260	280
45		202	225	247	270	292	315

We've worked this out for you based on 30 bushels per acre and a $7 per bushel price.

	50/50 Tenant-Landlord		3/5-2/5 Tenant-Landlord		2/3-1/3 Tenant-Landlord		3/4-1/4 Tenant-Landlord	
Income	$105.00	$105.00	$126.00	$84.00	$140.00	$70.00	$157.50	$52.50
Expenses								
Seed	5.00	5.00	6.00	4.00	6.66	3.34	7.50	2.50
Fertilizer & Lime	6.00	6.00	7.20	4.80	8.00	4.00	9.00	3.00
Chemicals	6.00	6.00	7.20	4.80	8.00	4.00	9.00	3.00
Machinery-Variable	10.00		10.00		10.00		10.00	
Crop drying, Misc.*	5.00	5.00	6.00	4.00	6.66	3.34	7.50	2.50
Total costs	$32.00	$22.00	$36.40	$17.60	$39.32	$14.68	$43.00	$11.00
Return	$73.00	$83.00	$89.60	$66.40	$100.68	$55.32	$114.50	$41.50

Miscellaneous covers interest, crop insurance, and grain handling.

The same guidelines will apply for soybeans as for corn when figuring which way to go. Supply and demand, quality of the tenant, and value of the land should always be considered.

Also, the same guidelines will apply to other crops you might grow.

There are two other areas to consider carefully in your crop share lease—cornstalks and marketing grain from shared storage.

How do the landlord and tenant share the return from renting stalk fields to a neighbor?

The tenant may assume that if he had livestock of his own to graze the stalk fields, he wouldn't pay additional rent. Therefore, he reasons that if he rents them to someone else, he should get all the rent.

The landlord probably doesn't see it that way. He sees that as the tenant getting a bigger share than he gets.

There's probably a bigger advantage to the tenant in grazing cornstalks than there is for the landlord. It reduces the volunteer corn the next year. Both face the disadvantage of loss of nutrients if the stalks are grazed too severely.

Probably the best answer in a crop share lease is for the rent paid by a third party to be split between the landlord and tenant the same as the crop is shared.

There should also be rules that the stalks are not to be grazed too intensively.

How do you market grain if the shares owned by landlord and tenant are stored in a common bin? Do they both have to sell at the same time?

Probably not. You can probably measure the bin and estimate fairly closely the amount each has in the bin.

The party selling first can then market up to 90% or even 100% of his estimated share—depending on your agreement. Then you can make your final settlement when both parties have sold all their grain from the bin.

For example, suppose you estimate that there are 10,000 bushels of grain in the bin and each party owns half. If the tenant wants to sell, he might be allowed to sell 90% of his estimated share—or 4500 bushels. If, when the landlord sells the rest, there are still 5700 bushels, he would settle with the tenant for his 600 bushels—the difference between half the actual grain stored (5100 bu.) and the amount the tenant sold earlier (4500 bu.).

Your lease can also spell out the price. Should it be the price the tenant received when he sold or the price the landlord received when he sold? That may be a bargaining point in the lease. But it should be put in writing when the lease agreement is made.

First rule to designing a good crop share lease is to push your pencil a lot. Use the guidelines here and apply your own figures to your own situation for the best results.

Other factors in your lease

It's tough to cover every detail that might come up during the lease period. But the more problems you anticipate, the smoother your working arrangement will be, whether you're the landlord or tenant. Think about these.

• Timeliness. One landlord reports his land was consistently the last of three farms his tenant operated when it came to planting and harvest. Result: yields were consistently about 10% lower than when he had farmed it himself.

On his 50% of the crop, that cost him about 5 bushels of corn per acre.

An agreement between landlord and tenant can specify that certain practices are to be done by a certain time. Of course, the time specified in the lease has to consider other land the tenant farms and be flexible when bad weather delays field work.

No landlord should always expect his crops to be the first planted—or the last. But he and the tenant should discuss the amount of land the tenant is farming, the adequacy of his equipment, and if he has enough labor to get the jobs done on time.

• Conservation practices. A farm that has existing grass waterways, terraces, and contours should be farmed in a manner to maintain them. A landlord who visits his farm and finds grass waterways plowed up may be quite upset.

As in a marriage, it is difficult to change the person once you enter into a contract with him. So a tenant who wants to improve the conservation of a farm had better make sure the landlord is willing to cooperate before he signs the lease.

• Chemicals and fertility. Goals of the landlord and tenant need to be similar when it comes to fertilizer and chemicals in a crop share lease. Discuss the fertility program in advance.

A tenant who is strong on heavy fertilizer applications may shock his landlord when he presents him with his half of the bill. One landlord was totally against use of any chemical. The tenant spent many extra hours and had a much higher machinery cost the year he farmed that land.

• Records. Some landlords are content just to get a bill from the tenant for their share of the expenses. Others want a detailed breakdown of costs plus reciepts that might show any discounts.

All these possible conflicts can be avoided by a thorough discussion before the lease agreement is made. Then they should all be put in writing so no one forgets. That can avoid a lot of conflicts.

Advantages of crop share

Crop share as compared to other type leases has some good potential for the landlord—and tenant, too.

In good crop years, yields are high so the landlord gets more than he likely would from cash rent.

There's more chance to supervise some of the farming practices. A normal cash rent arrangement leaves nearly all cropping decisions in the tenant's hands.

This participation can be enough to qualify the landlord for social security coverage if he wants to build up higher benefits. However, it's more common for a landlord to want to avoid too much participation if he's already retired. (See the chapter on material participation.)

A crop share lease may help the landlord qualify for a special use valuation that may trim the tax on his estate. But, again, that may conflict with his social security benefits.

The landlord gets to share in yield increasing technology with crop share. Choosing a tenant who uses yield builders such as narrow rows, a good fertility program, and effective chemicals will reap extra profit for the landlord.

Advantages for the tenant often make the crop share lease attractive for him, too.

While he may not come out quite as well in a high yield year as with cash rent, he doesn't suffer as much loss in a bad year.

The tenant's cash flow demands aren't as great with a crop share lease as with cash. He only has to stand the cost of about one-half the inputs as compared to all of them plus the cash rent with a cash lease.

A crop share lease may be best for a young farm family just starting to farm because of a lower risk. Some lenders won't even finance a beginning farmer with a cash lease unless he has a co-signer.

Disadvantages of crop share

Probably the biggest problem landlords face with a crop share lease is figuring out who pays what expenses. It's tough to determine what's right, especially as the economy changes.

It's also difficult to accurately predict your income. Yield and price greatly influence the amount of money you will have for living expenses. And it can vary greatly from year to year.

There's also the risk factor. You spend money to put the crop in with no guarantee on return. Plus, the landlord has more money tied up in the crop than with some other leases.

It takes some extra management with a crop share lease. You have to make the decision on crop insurance to cover your share of possible losses. You'll also have grain that calls for marketing know-how.

For the tenant, a crop share lease may limit decision making some. Since the landlord is paying part of the costs, he/she may want to have some say in the amount of fertilizer and chemicals used, for example.

The tenant also only gets part of the incresased return from new technology he uses. In other words, his return for gearing up for narrow rows won't pay off as well as if he were farming his own land or cash renting because the landlord will get half the yield increase.

Summary

A good crop share lease can offer both landlord and tenant a good way to get the most return from the land. But it does take time, patience, and effort to make sure all possible conflicts are anticipated and covered.

We suggest you also study the chapter on cash rent.

Even figure what a fair cash rent would be for the land. Then compare it with what the returns for both landlord and tenant are expected to be in an average year with crop share.

Remember that the tenant shouldn't expect to come out quite as well in a good or average year with crop share as with cash rent. The landlord will do a little better—and should—because he is sharing more risk.

Finally, judge the fairness of your crop share lease on the total agreement rather than on individual sections. Only the parties involved can make the final decision as to whether or not it is a fair lease. There are many bargaining areas.

Questions and Answers

Question: My tenant on a 50-50 crop share lease wants to plant oats to chop as silage. I don't have any livestock. What would be a fair split of the oat silage?

Answer: Since you don't have livestock to use the oatlage, you may want to cash rent for the value of one-third the oat crop and let the tenant pay all the production expenses. Another alternative is to take 50% of the oatlage if you have a definite buyer.

Still another alternative is to have the tenant leave a check strip to be combined to determine yield of the oats and straw. Then you should receive rent based on what it produces.

Question: I rent my small farm to a neighbor and we get along very well. I pay half of all the fertilizer and chemical costs, custom spraying, custom application of fertilizer, combining, drying, and trucking to a local elevator. Is this a fair agreement?

Answer: In most crop share leases, the tenant assumes all harvesting costs. The landlord may pay for normal shelling costs on his share of the corn. Landlords usually share only in custom spraying costs when it is an emergency, such as aerial spraying. However, the cost of the chemicals is normally shared. Otherwise, your lease is fairly typical.

You may want to do some additional pencil work to see if your lease arrangement still seems to be a fair one. Be sure to judge the lease in total as to its fairness rather than on specific items.

Question: My crop share lease is different than most. My tenant furnishes all the gas and labor plus half of the fertilizer, herbicide, and seed. I provide the land, storage, machinery, and I insure the crop until it is divided. How should the crop be divided?

Answer: You have several alternatives that you might want to consider. If this is high producing land, a division of 75% to the landlord and 25% to the tenant would probably be about right. On lower producing land, the tenant's share should be closer to 30%. Each party should then pay the drying costs on his share of the crop.

The biggest difference between this arrangement and a typical 50-50 crop share agreement is the machinery. You might consider dividing the grain equally and having the tenant pay rent on the machinery. Then, the crop insurance should be divided too.

Question: I rent land on a 50-50 crop share basis. As the tenant, I furnish the fuel, harvest the crops, and bale the hay. I have half interest in the heating unit on the bin dryer. What should the landlord's expense be on this set up?

Answer: Normally, the landlord would pay for his share of drying expense—fuel used in drying and cost of operating the dryer. In some cases, the landlord pays the tenant for normal field shelling costs on his share of the corn.

These may be some of the bargaining areas between the landlord and tenant in your situation.

Chapter 7

Cash Leases

Competition for land to rent has pushed cash rent upward. Question is, what's too high? Or, what is fair?

It's tough to decide on the right cash rent for farmland. The rent is set long before you know crop price or yield. The tenant has to live with that rent even if yields and prices leave him short of income to pay it. By the same token, if yield and price are better than expected, the landlord may find he would have made more money with a crop share lease.

Your first step is to decide if you really want to go the cash rent route. Take a look at the advantages and disadvantages.

Advantages

For the landlord:

• He/she is guaranteed a certain amount of income as long as the tenant is solvent and can pay. Cash rent takes away the risk of crop failure or a drop in price that might produce less income than was expected.

• The landlord is also relieved of making operating decisions. That can be a real advantage for the absentee or older landlord who doesn't want the responsibility of working closely with the tenant. The same goes for a widow or elderly landowner who may not be skilled in making farm management decisions.

• The landlord wouldn't be materially participating. So he wouldn't lose social security benefits. However, in some cases, he may want to materially participate to build his social security base or to qualify his estate for special use valuation.

• The landlord doesn't have to tie up money in production costs that aren't returned to him until after the crop is harvested as in a crop share agreement. In a crop share agreement he usually has to fork over a fair chunk of money for seed, chemicals, and fertilizer 6 months or longer before he has any crop to sell.

For the tenant:

• Many tenants like cash rent because it gives them full freedom in planning their farming operations. Maybe the tenant decides he wants to plant the whole farm to soybeans to eliminate some trips to the farm, for example. The cash rent agreement is likely to make it easier to make those decisions.

• The tenant gets full benefit from his top management abilities.

• The tenant takes more risk when he cash rents. His bet is that yield and price will be higher than the estimates he used to figure the cash rent. If they are, he gets full benefit from any yield and price increases.

Finally, for both landlord and tenant, the cash rent agreement reduces the risk of misunderstanding. As compared to a crop share lease that has to spell out who pays what production costs, the cash rent lease is quite simple. The renter's responsibility is to make the payment and farm responsibly. The landowner should provide the renter a good farming atmosphere.

Disadvantages

For the landlord:

• While cash rent cuts the landlord's risk, his return has historically been less than from a crop share lease.

• Cash rent can easily become too low in times of rising prices and increasing yields. In fact, with inflation, new terms— higher rent—may have to be negotiated more often than with a crop share lease. However, in some areas, tenants have bid up cash rents to the point where returns may be equal or higher for the landlord in a normal year with a cash rent lease than with a crop share lease.

• The tenant may not keep the farm up as well with cash rent. The tenant feels more free to use farming practices he chooses. He might try to stretch his acreage by plowing up some waterways, for example. Or, he may deplete the fertility and then quit farming the land.

For the tenant:

• The landlord may be reluctant to make needed farm improvements. For example, land improvements such as soil and water conservation practices won't increase his return the same way as they would with a crop share lease. With crop share, the landlord can see that he is going to get a share—usually 50%—of the increased yields from those practices. With cash rent, the only way he gets more return is to increase the rent.

The landlord probably isn't interested in putting up grain storage facilities in a cash rent arrangement because he won't have any grain of his own to store in them. Cash rental units are also less likely to have good livestock facilities.

• Probably the biggest drawback for a cash rent tenant is the higher risk he must handle. He can pile up a big investment by the time he pays all crop production costs plus the cash rent. Then, if the yields are low or the bottom drops out of the price, he gets hurt.

To cover this, some tenants get their landlords to agree to a disaster clause in the lease. Others make sure they have plenty of insurance on the crop. A new type lease—flexible cash rent—has become one popular way to ease this problem.

• If the tenant does a super job of farming and increases production of the farm significantly, the landlord is likely to say the rent should be higher. Landlords also tend to remember the high yield years better than the low ones.

These disadvantages point out that a cash lease should go further than just setting the rent figure. But, first, let's look at how you determine the dollar amount.

Ways to figure cash rent

The toughest job of all is to determine how much the cash rent should be. Both landlord and tenant want it to be a rate that's fair to each. There are basically five ways to determine a fair cash rent. They are:

(1) A percentage of the market value of the land.

(2) Landlord's expenses plus a return for his investment.

(3) Cash rent equal to the expected value of about one-third of the expected yield of the crop on good soils and less than one-third on poorer soils.

(4) A cash rent based on a typical crop share return.

(5) The going rate for cash rent in your area.

You may want to use one of these methods. But a combination of two or three of them will probably be a better way to be sure the rate is fair. Let's look at each of the methods separately.

A percentage of market value

Since the landlord could sell that farm and invest his money elsewhere at a good interest rate, many figure they ought to get a rent that's close to that. Of course, the tenant can argue that the landlord wouldn't have the full amount left to invest after paying Uncle Sam a sizeable chunk of tax on the gain. Plus, he would be giving up future land value increases due to inflation.

Historically, landlords and tenants using a percentage of market value of the land for cash rent have used a rate of 6% to 8%. But with rapidly rising land prices, you may have to adjust that down a little.

For example, for land that would sell for $3000 per acre, rent would be $180 to $240 using those percentages. By the time the tenant adds his production costs, there might be no way he could cover that much cash rent. It would take 72 to 96 bushels of $2.50 corn and 26 to 34 bushels of $7 soybeans just to cover the cash rent. Add another $110 for production costs and he'll need 44 more bushels of corn and 16 bushels of beans just to break even. Estimate your own costs. The $110 may be too low in some cases, high in others.

On top of that, the tenant will still need to cover his machinery investment and get a return for his labor.

Landlords can't expect a return from land equal to other fixed value investments.

So, while this method may work for some farms, inflation has made it tough to use it as the only way to figure a fair cash rent.

Expenses plus return on investment

Landlords still have certain expenses even when they cash rent their land. These include taxes, insurance, depreciation, and upkeep.

The first step in this figuring is to add up all those costs. Then use a percentage of the owner's original investment in the land on top of that.

For example, suppose taxes, insurance, depreciation, and upkeep add up to $20 per acre. Say he originally paid $1200 per acre for that land and wants a 7% return. That's another $84. So the cash rent might be set at $104 per acre.

Again, this method can be misleading. In fact it won't work for everyone. If the land was bought many years ago for $200 per acre and it's now worth $2000 per acre, even an 8% to 10% return on that original investment isn't going to give an accurate value for cash rent. Then, you're going to want to look at some of the other methods for estimating cash rent.

Value of expected yield

With this method, the landlord often receives a cash rent that is equal to what the landlord and tenant figure one-third of the expected yield of the crop will be worth.

For example, suppose you figure the corn yield will be 120 bushels per acre in an average year. The landlord would then get a cash rent equal to what 40 bushels of corn will be worth. If you figure corn will be worth $2.50 per bushel, cash rent will then be set at $100 per acre. If soybeans average 42 bushels per acre and you expect a $7 price, $98 per acre cash rent would be about right with this formula.

You may have to make some adjustments for different crops grown.

Share of the crop method

Basing cash rent on the value of one-third of the expected crop and price is probably the easiest way to figure cash rent.

Simply estimate the average crop yield from past experience. Then you multiply that by the estimated price. You then figure one-third of that as the cash rent. One third is the normal rate on good soils. You may have to go with a lower percentage on poorer soils.

An example: Say corn yield will average 120 bushels per acre and you predict a $2.50 per bushel price.

Gross return per acre will then be $300. So cash rent might be set at $100 per acre (1/3 of $300).

Another example: Say soybean yield will average 42 bushels per acre and you predict a $7 per bushel price.

Gross return per acre will then be $294. So cash rent might be set at $98 per acre.

If you figure cash rent this way, you'll find that it usually comes out that the landlord gets slightly less than he would with a crop share lease. But, he avoids the risk he would have with crop share so he has to expect a little less.

Biggest trick is to predict both yield and price as accurately as possible. Look at past performance on yield. Use the most reliable forecasts you can find on price.

Crop share return

This method may be the most accurate way to figure a cash rent that's fair. You estimate all costs the landlord would have under a crop share lease. Then you estimate yield and price to see what return the landlord would normally receive under a crop share lease. That gives a rough guideline for what the cash rent should be.

However, you need to drop back from that a bit. Figure that the renter is assuming all the risk in a cash lease.

The tables here show growing cost estimates for corn and soybeans. They also show returns at different yield and price levels after those costs have been subtracted. That's what the tenant will have left to cover machinery ownership, labor inputs, cash rent, and profit.

Variable cost estimates of corn per acre

	Yields	80 bu.	90 bu.	100 bu.	110 bu.	120 bu.	130 bu.
Seed		$10	$11	$11	$12	$12	$12
Fertilzer & lime		24	26	29	32	36	40
Herbicide & insecticide		13	14	15	15	15	16
Machinery var. costs		17	17	18	19	20	21
Crop drying, ins., misc. & int.		18	19	20	21	22	23
Total Variable Costs		$82	$87	$93	$99	$105	$112

Return to tenant

Corn yield (bu.)	Corn price				
	$2.00	$2.25	$2.50	$2.75	$3.00
80	$78	$98	$118	$138	$158
90	93	116	138	160	183
100	107	132	157	182	207
110	121	149	176	204	231
120	135	165	195	225	255
130	147	180	213	245	278

Variable cost estimates of soybeans per acre

Yield	25 bu.	30 bu.	35 bu.	40 bu.	45 bu.
Seed	$ 9	$10	$10	$11	$11
Fertilzer & lime	11	12	13	14	15
Chemicals	11	12	13	14	15
Machinery var. costs	10	10	11	11	12
Misc. int., crop ins., etc.	10	10	10	10	10
Total Variable Costs	$51	$54	$57	$60	$63

Return to tenant

Bean yield (bu.)	Soybean price				
	$5.00	$5.50	$6.00	$6.50	$7.00
25	$ 74.00	$ 86.50	$ 99.00	$111.50	$124.00
30	96.00	111.00	126.00	141.00	156.00
35	118.00	135.50	153.00	170.50	188.00
40	140.00	160.00	180.00	200.00	220.00
45	162.00	184.50	207.00	229.50	252.00

Let's look at an example. Suppose the farm being considered will produce 120 bushels of corn and 40 bushels of soybeans per acre in a normal year. Say you expect corn price to be $2.50 and soybeans $7.

Here's how the landlord's returns would shake out for corn:

1/2 of yield	**60 bushels**
Expected Price	$2.50
Gross return	$150.00
Expenses	$42.50
Net return	$107.50

Here are the estimates for the landlord's return for soybeans:

1/2 of yield	**20 bushels**
Expected price	$7.00
Gross return	$140.00
Expenses	$24.50
Net return	$115.50

So the average net return for the landlord figures out to be about $111.50 per acre if corn and soybean acreage are equal.

He will still have taxes, insurance, and upkeep to pay out of that. If those total $20 per acre, he would expect to net about $91.50 per acre with a crop share lease. He will want to approach that same net return with cash rent.

However, remember that the tenant is taking all the risk with cash rent. If yield or price drops from those estimates, he can lose. The landlord reduces his risk with a cash rent lease. So even if price or yield drops, he gets his money.

That means the cash rent is often dropped some from what the landlord might expect to get from a crop share lease. A 5% to 10% risk factor is often used.

So in our example, the cash rent might be dropped about

10% to about $100 ($111.50 × 90%). The $11.50 reduction is the break the tenant gets for shouldering all the risk.

Use your own figures for these calculations if they're available. If not, the tables here provide good guidelines.

Now, let's look at this just from the tenant's view. The return to tenant section of each table shown earlier shows the amount of money the tenant will have after paying the variable costs. That money has to cover the cash rent, his machinery investment, and labor costs to grow and harvest the crop.

An example: Suppose you have to use $30 per acre to make machinery payments and need $25 per acre for living expenses. Say you want to cash rent land that averages 100 bushels of corn and you expect price to be $2.50 per bushel when you sell. What's the top cash rent you can pay?

From the table on corn, you can see that you can expect $157 after variable costs for 100 bushels of $2.50 corn. Then, subtract the $55 cost you have to make for machinery and labor, and you find the top cash rent you can afford to pay is $102 per acre.

But since you take all the yield and price risk, you would probably want to back that down a bit. You might apply the 10% risk factor to reduce the rent to no more than $92 per acre ($102 × 90%).

Going rate for cash rent

Before you finally set the cash rent, check the rate on neighboring farms that are similar to the one you're working with. By checking a number of these, you can probably get an average cash rent that is being paid for similar land in your area.

Summary

There's no one perfect way to determine what's fair in cash rent. A combination of two or more of these five methods will probably be your best bet. Only the parties involved can make the final determination of what is fair. But these methods will certainly help.

Questions and Answers

Question: I have 200 acres to rent out. I've heard that some land is rented out at auction to the highest bidder. This is new to me. Are there any pitfalls or advantages?

Answer: Yes, you lose control in selecting a tenant and limit your control of the care and conservation that is given your land if you use a rental auction. You may want to specify a maximum acreage of row crops to be grown. Of course, that may lower the bids.

You can accept written bids. But we suggest you reserve the right to accept or reject any or all bids.

However, if you are anxious to receive the very highest return possible, the auction method would probably be the way to go. Then, you could contact your local auctioneers for details.

Question: I rent my farm for cash. I have only corn, soybeans, and some alfalfa. Should the alfalfa land be rented at the same rate as for other crops?

Answer: Hay land normally rents for less than row crop land on a cash basis, especially if hay acreage is part of a long term land use program. If alfalfa is grown at the tenant's request, cash rent may be nearly the same as for row crop.

Question: I have rented farmland on a 50-50 basis for the last seven years. Now I want to cash rent or get a percentage of the crops instead—like one-third of the corn and two-fifths of the beans. I provide a home and buildings for the renter. What would be a fair way to do this?

Answer: Cash rent would be the simplest way. You would include a rent for the land plus the facilities. The tenant would then pay all operating expenses.

However, if you prefer a crop percentage base with the tenant paying all operating expenses, a rental charge may be negotiated for the facilities, assuming they are modern and well maintained. The percentage of the crop lease might stipulate one-third of the corn and one-third to two-fifths of the beans. Pencil it out to see how close it comes to your present 50-50 crop share lease. But remember that you may have to take a little less on a cash basis since you no longer shoulder any of the risk.

Question: I cash rent my 300 acre farm on a five year contract. But because of inflation, we are not getting enough rent at the end of the contract. What suggestions do you have?

Answer: Consider basing the rent on a percent of the crop produced times the market price. Another alternative is to establish a minimum cash rent plus an adjustment factor based on yield and price. Or, you might go to a shorter term lease or put an adjustment factor in the lease. Long term cash leases don't work very well in today's rapidly changing economy.

We suggest you take a look at the chapter in this book on flexible cash rents.

Chapter 8

Flexible Cash Rent

Conflict between landlords and tenants about the amount of cash rent that should be paid has given birth to a fairly new kind of rent. Flexible (also called variable) cash rent shifts some of the risk away from the tenant in poor years, but it gives the landlord a chance to make it up in the good years.

In a typical flexible cash lease, there's a base rent per acre based on low price and low yield. Then the final cash rent increases if price, yield, or both, end up above that base.

For example the rent might be $80 per acre based on 80 bushels of corn per acre and a $2 corn price—all expected to be low. The landlord might then get an additional cash rent per acre of $4 for each 10¢ price is above $2 per bushel on November 1 and an additional 25¢ per acre for each bushel yield goes over 80 bushels per acre.

Suppose corn yield was 100 bushels per acre and price was $2.20 per bushel on November 1. Yield would give an extra $5 per acre (20 bu. × 25¢) and price would increase the cash rent by $8 per acre ($4 × 2). So the final cash rent would be $93 per acre ($80 + $5 + $8).

If the yield or price rise, the cash rent also rises. So when prices and yields are up both landlord and tenant share in the bounty. But when the profit picture is bleak, both tighten their belts. It's kind of a combination cash rent and crop share lease.

There are several different flexible rent methods being used. They're described later in this chapter.

But first, take a look at some of the advantages a flexible cash lease might offer.

• It's not necessary to negotiate over the amount of cash rent each year. If the final rent is tied to yield and price, that will keep the rent increasing as price increases.

• Since the landlord does benefit from higher yields, he may be more willing to make land improvements.

• Landlord-tenant relations are improved because there are fewer decisions and agreements as compared to some other rental agreements.

• A flexible cash lease may encourage livestock production more than a straight cash or crop share lease. Both parties may consider feeding the grain through livestock when grain profits are down.

• A tenant's ability to pay cash rent is usually related to the level of prices. So his risk is eased. Plus, he's better able to pay a higher rent in an especially good year.

• The flexible lease is especially handy for an absentee owner or a local owner who likes the crop share lease, but doesn't want to be bothered with making farm production and marketing decisions. The landlord will likely have less chance to materially participate with a flexible cash lease than with a crop share lease. And this could affect social security payments.

The base rent

Your first step is to figure how much the base amount should be. There's no set figure that will apply to everyone. But there are some guidelines.

Normally, this amount is low in comparison to what the landlord might expect to get from a regular cash or crop share lease. But it should be an amount the landlord feels he can live with in case a bad year leaves him at that amount.

The base rent should cover taxes, insurance, and a little return for the landlord's investment. If the land has to produce enough for the landlord's living expenses, figure enough to at least cover the minimum he will need.

For example, the rate might be set at taxes, plus insurance, plus a 3% return on the land value. Suppose taxes are $15 per acre, insurance $1 per acre, and the land is worth $2000 per acre. The base rent would be about $76 per arce.

Another method might figure 20% to 25% of the value of an average crop. Say the land normally produces 100 bushels of corn, for example, and you expect price to be $2.40 per bushel. Total value would be expected to be $240. So you might set the base rent in the $48 to $60 per acre range.

Still another method for setting the base price for corn land is to estimate the average yield, multiply it by $1, and figure 35% of that as the base rent. However, some landlords would consider this a low base.

For example, suppose average corn yield is 120 bushels. The rate would be 120 × $1 × 35% = $42 per acre base rent.

For soybeans, the dollar figure for the base rent might equal the average yield—40 bushels per acre equals $40 per acre base rent.

Final rent goal

Setting the base rent too low puts more importance on the adjustment rates. Doing a good job of setting the adjusted rates becomes more critical, so the final rent is in line with the real rental value of the land.

Both landlord and tenant hope the final rent will be somewhat higher than the base rent.

The formula will usually be set up so the landlord gets nearly as much return in an average yield and price year as he would get with a crop share lease.

Kinds of flexible leases

There are many kinds of variations in flexible leases. But they basically narrow down to a base rent plus one of the following:

(1) Additional payment based on price.

(2) Additional payment based on price and yield.

(3) Additional payment based on the value of so many bushels of the actual crop.

(4) Additional payment based on the value of a percentage of the actual yield.

Price method

The additional payment based on price method sets a base cash rent. The base rent could be the value of a certain number of bushels of production. The tenant pays that amount even if the crop is a complete failure.

They also determine a low price to start with. In many cases, that has been $1.50 to $2 per bushel for corn. They, of course, expect price to be higher on the date they set for making the final determination.

Then they figure a dollar amount per acre that cash rent will increase for each 10¢ increase in corn price above $1.50 on the target date, for instance.

For example: Suppose base rent is $70 per acre plus $4 per acre for each 10¢ corn price is above $2 on December 1.

Suppose corn price is $2.30 per bushel on December 1. Then final cash rent will be $82 per acre. That's the $70 base rent plus $12 ($4 × 3). The 3 factor comes from three 10¢ increases above the $2 base price.

This formula will work both ways on price. It can decrease the base rent if price is lower than the base. In the example above, if price fell to $1.60 per bushel, the total cash rent might drop to $54 per acre.

But the landlord may want to start with a higher base, then, to avoid a too low return.

One drawback for this system is that the tenant still shoulders a`lot of risk. Say he has a yield failure and price climbs. He might end up with a 50 bushel corn yield and $3 per bushel price. Then he would pay the $70 base rent + $4 × 10 = $40, or a total of $110 per acre cash rent.

Since he would only get $150 return, the tenant would be hurt.

For soybeans, similar figuring would be used. Say you set a $70 base rent plus $3 for each 10¢ bean price is above $6 on December 1. A $6.50 December 1 price would increase the rent by $15 to a total of $85.

You'll have to determine your own base figures. For example, say you set base rent at $35 per acre and base the increase on so many dollars for each 10¢ price is above $1.50 for corn on December 1. Suppose you figure a fair rent is $95 per acre in an average year and expect corn price to be $2.50 on December 1. You'll need to figure a $6 per acre increase for each 10¢ price increase to reach that final amount.

You can also use this method for other crops. Simply figure roughly what the cash rent will be in an average year. Then aim your formula to hit that if you do have an average year.

Price and yield method

A way to avoid the disaster situation described earlier for the tenant is to use both price and yield to determine the final rent. The rental agreement can be arranged so the final cash rent would go down if market price and/or yield drops, too.

The price and yield rental agreement sets a base cash rent per acre as well as a base price and yield for the crop.

For example, suppose you set the base rent at $60 per are. You might set the price change at $4 for each 10¢ price is above or below $2 on December 1. The yield figure might be a 25¢ per bushel change for each bushel yield is above or below 80 bushels.

If corn price is $2.45 on December 1, here's how that would affect the rent.

Crop price $2.45
Base price   2.00
Change $ .45
(45 ÷ 10 = 4.5 × 4 = $18)

Therefore, the 45¢ increase in corn price above the $2 base increases the per acre rent by $18—to $78 per acre. Here's how a 100 bushel yield would affect the rent.

Crop yield   100 bu./acre
Base yield   80 bu./acre
Change   20 bu./acre
(20 × 25¢ = $5)

So the total cash rent would be the $60 base, plus $18 increase from the higher price, plus the $5 increase from the higher yield for a total of $83 per acre.

With a 50-50 crop share arrangement, a landlord would expect to net about $85 per acre if price is $2.45 per bushel of corn that yields 100 bushels per acre.

So since the tenant stands more risk with even the flexible lease, the $83 figure in this example would be a fair final rent.

It's wise to compare the return for the landlord at various yield and price levels in your flexible rent with that he would net with those yields and prices in a 50-50 crop share lease. Generally, his return should be only slightly less at average yields and prices. There will usually be a wider spread in favor of the tenant as yield and price increase.

Price per bushel harvested

A landlord who wants to reap the benefit of increasing yield might like to figure the cash rent on a certain amount of money for each bushel of grain harvested. He won't get the benefit from an increase in price.

Here's how this method can work. You estimate the average yield of the crop and what would be a fair cash rent in an average year.

Then you divide the rent figure by the average yield to find the amount per bushel.

For example, say the average corn yield is 120 bushels and the cash rent figures out at $90. That's 75¢ cash rent per bushel harvested ($90 ÷ 120 bu.).

If yield is 130 bushels, the landlord will get $97.50 cash rent (75¢ × 130 bu.). If it's 100 bushels, he would get $75 total per acre (75¢ × 100 bu.).

For soybeans that average 40 bushels per acre on that same land, the rent per bushel might be $2.25 ($90 ÷ 40 bu.). So if yield is 45 bushels, the total rent would be $101.25 per acre ($2.25 × 45 bu.).

This method can also use a base rent. Say base rent is $50 in our soybean example. To get the extra $40 rent in an average year, the rent per bushel would need to be set at $1. But then if yield hits 45 bushels per acre, the landlord would only get an extra $5 for the extra 5 bushels yield.

Bushels of crop

Some landlords are willing to share price risk. It's a simple method in that the landlord takes his cash rent as the value of a certain number of bushels of the crop per acre.

There can be a base rent as with other methods—then a date and market quotation to set the price per bushel. Or if the landlord wants the marketing challenge, he may even take the actual grain as in a crop share lease. Then he determines when to sell.

For example, if the landlord figures corn prices will be about $2.50 per bushel, he might want 35 bushels per acre to cover his cash rent (35 bu. × $2.50 = $87.50).

If corn values slide to only $2 per bushel, the landlord receives only $70 rent per acre. However, if prices rise to $3 a bushel, then the landlord receives $105 per acre rent, thereby sharing in the bonanza.

With a $50 base rent in this example, the landlord might also get 15 bushels of corn. So at the $2.50 price, his total rent would again be $87.50. At $2, he would total $80. At $3, he would get $95 total cash rent.

Under this arrangement, the landlord assumes no production costs or yield risks.

Percentage of the crop

Another flexible arrangement option is where the landlord and tenant share both price and yield risks by agreeing to divide the actual crop harvested on a percentage basis. The percentage is normally based on each person's contributions.

For example, after evaluating input costs, the tenant might figure he contributes 70% of the enterprise and the landlord 30%. They may then agree on a 70-30 basis. For corn and soybeans, a 30% to 35% share for the landlord is usually considered fair.

If the yield turns out to be 100 bushels of corn per acre, and corn is selling at $2.50 a bushel, a 33% rent amounts to $82.50 per acre. But if corn prices slip to $2 per bushel, the rent would be $66 per acre. At $3, the landlord would receive $99 total cash rent.

Bigger yield will also push up the total cash rent with this arrangement. With a 120 bushel yield and a 33% share for the landlord, he will get a rent equal to the value of 39.6 bushels of corn. If corn is selling at $2.50, his total cash rent will be $99 (39.6 bu. × $2.50).

This arrangement could also use a base rent with a smaller percentage of the crop.

For example, say the base rent is $45, average yield is 100 bushels of corn per acre, and you expect price on your target date to be $2.50 per bushel.

A 15% share of yield would give a total rent of $82.50 per acre if yield is 100 bushels per acre and price is $2.50 per bushel (15 bu. × $2.50 = $37.50 + $45 = $82.50).

Other things to consider

A flexible lease adds many possibilities for a fair amount of cash rent. It may cut the tenant's take in especially good crop years. But it greatly reduces his risk in the bad years.

Its fairly easy to determine the lease arrangements. But there are some things to watch for—as in any lease.

Set the date and market quotation to be used if grain price is to be a factor in setting the final rental rate.

A November 1 or December 1 date is often used. Or you may want to use an average over several weeks. Also set a location and grade for setting the price.

Some farmers use a local price quotation. Others use a state average price as reported by the Crop and Livestock Reporting Service in their state. An average price for several weeks in November or December is often used.

Most flexible leases call for the crop to be weighed and moisture tested to determine yield. Be sure to agree on these arrangements—weighing practices and moisture content—in your lease.

If the landlord wants to receive a rent equal to one-third of the crop, it may be feasible to divide the crop 1/3—2/3 at harvest. Then the landlord may sell his/her share whenever he/she desires.

Be sure the arrangement provides incentive for the tenant to shoot for high yields. If the lease is too favorable to the landlord on yield increases, the tenant may put his emphasis on higher yields on other land he farms rather than the flexible lease land.

Also make sure the agreement promotes good conservation practices.

The flexible cash lease generally doesn't lend itself to a rental agreement on forage crops, pasture, or buildings. A cash rent agreement will probably work better on those. However, there are ways to use a type of flexible rent for pastures. They're described in the chapter about pasture rent in this book.

Estimate your own production costs closely before you negotiate on your flexible cash rent lease. The chapters on crop share and cash rent in this book will help. That will give you a good clue on what figures to use in your lease.

Summary

One major purpose of a flexible lease is to reduce the tenant's risk. In exchange, the landlord gets to share in some added returns if the yield or economic outlook turns around.

There are a number of good ways to build a fair flexible lease. The ones described here should give you a good start.

Chapter 9

Custom Farming, An Alternative To Leasing

For the landowner who is willing to take charge of the management of his land and assume all risks, custom farming may be an alternative to leasing.

This can also open the door for an operator who is looking for extra work for his equipment. Maybe he can't find enough land to cash or crop share rent. Or maybe he doesn't have the capital to farm more land, but has the labor and machinery capacity. Providing custom service can generate extra income without the risk of renting additional land.

How it works

Custom farming works this way: The landowner keeps the entire operation under his direct supervision. He decides how much fertilizer to apply, what to plant, and what herbicides to apply, for example. In other words, the landowner calls all the shots.

But he hires most or all of the labor and equipment. He may hire one or more farmers—often a neighbor looking for extra work for himself and his machinery—to do the various field operations.

Some advantages

Custom farming may be more profitable for the landowner than leasing out farmland. If the landowner has and wants to use his management ability but lacks machinery, it offers possibilities. Profits can be 20% to 40% higher when comparing custom farming to crop sharing. But the risks are also greater.

For some people buying expensive land, this extra payment is really needed.

For the custom operator, there may be several reasons to go this route.

Custom farming appeals to the operator who:

(1) Needs additional acres.

(2) Is short on capital.

(3) Prefers a guaranteed income and less risk.

In many areas, there is a strong demand for land. And, while the operator might prefer to rent on a cash or share basis, no land is available. So his alternative may be to hire himself out as a custom farmer.

Custom operators who have unused machinery capacity and labor must cover out of pocket costs, taxes, maintenance, and additional wear. Anything else is return for labor and recovering some, and hopefully all, of his fixed costs. This money may be available for debt repayment, living expenses, and to improve his financial position.

Finally, a good job of custom farming can help a young farmer become established in the community. It could result in his finding land on a cash or share rental basis—possibly land he has been custom farming.

Some disadvantages

The potential of higher income is offset by several drawbacks to custom farming for the landowner.

• Custom farming requires a lot more management time for the landowner. In fact, it can be a constant burden. You need to do all the jobs you take for granted your tenant will do——or hire them done.

For instance, you have to check the insects, disease, and weed problems; make sure the crop is dry enough to harvest;

and then inspect grain weekly while it's in storage. Plus, you must line up operators to do the various field operations. And there are still the weeds to mow and fences to repair.

• Because of that extra management you provide, custom farming will likely reduce or eliminate your social security benefits if you're retired. You would be materially participating—and that can cause some retired people to lose some or all of their social security benefits. In fact, your earnings would be considered self employment income—earnings from your management—and be subject to social security tax.

Cash or crop share rent income doesn't normally affect social security benefits or tax. The only exception is if you provide a lot of labor or management so that you are considered to be materially participating.

Management ability has to be top notch, too. That means the landowner has to read, study, and attend meetings to keep up to date on fertility, chemical use, and hybrids, for example. Marketing management may be the toughest.

The landowner also has to be able to manage people well.

• The landowner also assumes all the risk with a custom arrangement. He suffers the loss should there be a crop failure or low crop prices. The loss is more than with a crop share or cash lease because the landowner has spent money for the crop production plus the hired custom work.

• The landowner must furnish operating capital in much larger amounts than are necessary under share or cash rental arrangements. And cash must be available to pay for the hired work throughout the year.

• A major problem may also develop with untimely fieldwork. If that happens, yields are usually hurt. In other words, custom operators may do their own work first—and yours comes last.

What to pay

The going local rates are most often the guide for custom work. But be sure it's a fair rate and not a "doing it as a favor" rate. Other people rely on custom rate figures published by their state agricultural colleges. Here's a guide.

Custom Farming Rates

Tillage	Average	Range
Chopping cornstalks, per acre	$ 4.10	$ 3.30- 4.90
Plowing—moldboard, per acre	9.30	7.60-11.00
—chisel, per acre	7.60	6.35- 8.85
—disc (heavy), per acre	6.50	5.25- 7.75
Discing—tandem, per acre	4.90	3.80- 6.00
Harrowing, per acre	2.70	1.90- 3.50
Field cultivating, per acre	4.50	3.50- 5.50
Hoeing—rotary, per acre	2.90	2.10- 3.70
Cultivating, per acre	4.00	3.00- 5.00

Planting	Average	Range
Planter—with attachments, per acre	$ 6.20	$ 4.70- 7.70
—without attachments, per acre	5.35	4.30- 6.40
Till planter, per acre	9.50	7.00-12.00
Grain drill, per acre	5.15	3.65- 6.65

Fertilizer Application (materials not included)	Average	Range
Dry bulk—applied, per acre	2.35	1.75- 2.95
—applicator only, per acre	1.00	.50- 1.50
Liquid—applied, per acre	3.00	2.35- 3.65
—applicator only, per acre	1.65	.65- 2.65
Gaseous—applied, per acre	4.35	3.25- 5.45
—applicator only, per acre	1.30	.60- 2.00

Spraying (materials not included)	Average	Range
Ground, per acre	3.05	2.25- 3.85
Aerial, per acre	3.70	2.80- 4.60
Ground, incorporated, per acre	6.00	5.00- 7.00

Harvesting Grain	Average	Range
Picking ear corn, per acre	14.00	11.00-17.00
Picker sheller, per acre	17.00	14.40-19.60
Corn combining, per acre	20.60	18.25-22.95
Corn drying, per bushel	.18	.11- .25
per % moisture removed	.030	.025- .035
Corn shelling, per bushel	.085	.055- .115
Soybean combining, per acre	18.65	16.00-21.30
Small grain combining, per acre	16.20	13.70-18.70

Harvesting Forages

Hay—mowing, per acre	3.70	2.90- 4.50
—conditioning, per acre	4.50	3.00- 6.00
—raking, per acre	2.85	1.90- 3.80
—combination of above, per acre	6.90	5.30- 8.50
Hay baling—square, per bale	.27	.21- .33
—large round, per bale	5.75	4.85- 6.65
—stack, 1 ton, per stack	10.30	8.70-11.90
—stack, 3 ton, per stack	22.00	16.30-27.70
Silage—harvesting, per acre	36.00	26.00-46.00
—harvesting, per hour	50.00	39.00-61.00
—harvesting, hauling & blowing, per hour	60.75	37.25-84.25

Tractor Rental (operator and fuel not included)

per horsepower, per hour	.115	.10- .13
per horsepower, per day	1.10	.90- 1.30

Custom Farming—growing and harvesting

Corn, per acre	69.50	58.00-80.00*
Soybeans, per acre	62.40	52.80-72.00*
Small grain, per acre	39.00	29.00-49.00*

*Custom farming rates vary because of differences in custom work performed and whether or not an incentive plan is included.

These custom rate guidelines will help you set the rate. But you'll be wise to compare them to the going rates for custom work in your own area. Also keep in mind that these are present guidelines and will become out of date. So update them each year.

The rates, unless otherwise noted, include costs of machine ownership (interest, depreciation, and maintenance), the power unit used, fuel costs, operator labor cost, total overhead, and profit.

Also consider availability of machinery in your area, timeliness, operator skill, and the performance characteristics of the machine being used. These can affect the rate.

Note that the total custom farming rates at the bottom of the table give a wide range. That's because of differences in custom work performed and whether or not an incentive plan is included. It may be better to add the individual operations to get your total.

An example: For your particular operation, suppose you add the following operations.

Moldboard plowing	$ 9.30
Discing (2 times)	9.80
Harrowing (2 times)	5.40
Planting	6.20
Dry bulk fertilizer	2.35
Gaseous fertilizer	4.35
Rotary hoeing	2.90
Cultivating	4.00
Spraying	3.05
Harvesting (corn combine)	20.60
Total	$67.95

With those operations, it might cost you roughly $68 an acre to get all the jobs done by a custom operator.

Landowner's returns

With custom farming the landowner gets all the crop to sell. So if yield is 110 bushels of corn and price is $2.25, he will get a $248 return.

In addition to the $68 custom cost in the example figured earlier, the landowner has other costs. For a 110 bushel yield, these are estimated at about $80 (seed—$12; fertilizer—$32; herbicide and insecticide—$15; and crop drying, insurance, interest, and miscellaneous—$21).

Therefore, his total cost will approach $148. That leaves $100 net return—before taxes and upkeep.

On a crop share basis, this farmer—with the same yield and price—could expect to net about $84. So his net would be $16 per acre more in this example, or 19% more than he might expect with a crop share lease.

But don't forget that it takes a lot more skill from the landowner to manage the farming operation. And if timeliness of operations suffers enough to deteriorate yields, his return can take a sharp cut. Or, if price or yield drops severely, he gets hit hard because of his high investment per acre. In other words, there's more risk.

Plus, as the owner, there are other costs such as hauling supplies to the farm and crops to market to be considered.

How to hire

One of two approaches can be followed to hire custom farming done. Some landowners prefer to hire individual jobs done. Others like to use an incentive agreement with one custom operator to do all the jobs involved with producing the crop. Such an agreement would guarantee a flat cash payment per acre plus a bonus. That should stimulate the custom operator's interest to get the work done on time.

The incentive plan

The incentive payment, based on yield, rewards the custom operator for timely plowing, planting, and harvesting. If based on yield, the base yield should be low enough that the operator would likely be paid an incentive 9 out of 10 years. So you might set the base at 10 bushels below the average yield, for example.

One landowner paid the custom operator $50 an acre for all fieldwork from plowing to the time the crop was stored. As a reward for doing a good job—getting increased yields—the custom operator was paid an incentive of one-third the yield over 90 bushels per acre on corn and 32 bushels per acre on soybeans—average yields.

That proved to be a good bonus with corn for the custom operator. Corn averaged 141 bushels an acre on this farm that year. The custom operator thus received a bonus of 17 bushels per acre—one-third of the 51 bushel yield over 90 bushels per acre. That would mean $39.10 an acre incentive with corn at $2.30 a bushel.

Soybeans on this farm yielded 35 bushels an acre. The custom operator's one-third share over 32 bushels per acre was only one bushel, or about $7 an acre additional income.

Of course, an incentive program will work only if the same operator does all the field operations.

The landowner and operator should agree on responsibilities before entering into an incentive agreement. How and when the operator will apply herbicides should be clearly stated. Fertilizer can be custom or operator applied, depending on weather and availability.

Custom farming can work. Just make sure the agreement with the custom operator covers all the details such as timeliness and all the other things he is expected to do.

Chapter 10

Livestock Share Lease

A typical livestock share lease is a close relative to a crop share lease. Landlord and tenant each provide a share of the inputs. Then returns are normally shared in that same proportion.

The livestock share agreement normally includes a total farm operation—crops plus a livestock program. Let's first assume you're aiming for a 50-50 split on both contributions and returns. That's probably the most typical arrangement. However, we'll also look at other splits later in this chapter.

On the crop production, the tenant normally provides his labor and machinery plus half the production costs, such as seed, chemicals, and fertilizer—like a typical 50-50 crop share lease.

The landlord provides the land, pays the taxes and insurance, and furnishes half the production inputs—again, like a 50-50 crop share lease. In fact, you can probably follow the guidelines in the chapter about crop share leases and be right on target.

There are usually several big differences, though. The landlord often furnishes half the fuel for the total farm operation, including crop production. But he may not pay any fuel on grain not marketed through livestock. He may also pay half or part of the cost of harvesting silage and baling hay in a cattle operation.

In the livestock end of the lease, the tenant provides his labor, half the production costs such as feed and health care, and half the livestock.

The landlord normally furnishes the livestock facilities and part or all of the equipment, half the production costs, and half the livestock.

Then returns are split evenly between the landlord and tenant.

Let's take a closer look at some details of who provides what in a typical livestock share lease.

The owner furnishes materials for repairs. He also pays for the labor for remodeling and major repairs on his property, such as buildings and equipment. The tenant usually provides the labor for minor repairs. Be sure the lease spells out what a minor—or major—repair is.

In other words, the landlord would likely furnish his own or pay for the labor for building new fences or a building remodeling or construction project. The tenant handles the labor for fence repairs or temporary fences and minor building repairs. The landlord usually provides the materials for these. But these may be good areas for some give and take bargaining.

Seed, fertilizer, and chemical costs are shared in the same proportion as the crop is shared. In the livestock operation, feed, bedding, salt, minerals, veterinary fees, marketing costs, and breeding fees are shared in the same proportion as the livestock sale proceeds are shared.

Typically, you can follow the same guidelines in splitting these costs as with a crop share lease. That chapter in this book may give you some additional ideas.

Is it fair?

The big question is whether or not the 50-50 split is fair. Are the landlord and tenant each actually contributing 50% of the total input?

You can put a pencil to your arrangement and get a good idea. But it's still tough because you have to decide what interest rates to use and how much the labor contribution is worth. Those things can float all over the board. Here are some guidelines:

- For land, you can use a figure fairly close to its fair market value, then take a percentage return to figure the landlord's contribution. Most economists suggest you use a 3% to 5% return—about what most owners expect to get from their land in a rental agreement.

Then add actual taxes to round out the land contribution.

- Buildings and equipment can be valued with the land or figured separately. If you include these with the land, you may want to boost the percentage return by about 1/2%—5% rather than 4 1/2%, for example.

If you list buildings and equipment separately as a contribution, 5% to 10% of their value will usually give you a good guideline as to the owner's annual contribution. Consider useful life. Then divide that life into 100 to get the percentage. A 20 year life, for example, would give a 5% rate (100 ÷ 20). Then include interest, repairs, property tax, and insurance costs. (The chapter on building rent will give you some more ideas.)

- Each party should list his expected contribution toward labor and management. This may be based on an hourly rate times the number of hours each expects to work.

Try to be realistic. No one takes on the hard work and risk of farming for an expected return of $10,000 or less any more. Besides, the operator is also providing his management skills.

You might figure the rate in line with what a laborer or factory worker in your area makes, then add at least several thousand dollars for management skills.

- Livestock contribution is often figured at 8% to 10% of each person's investment in the livestock. In a 50-50 arrangement, this should be the same for both landlord and tenant.

- Machinery cost can be based on actual depreciation of the machinery. Or you may want to estimate it at $40 to $50 per acre. Probably the best way is to take original cost of each piece of machinery, subtract salvage value, and divide that by its estimated useful life. That gives your cost per year.

Let's set up a hypothetical example for a livestock share lease that might be quite typical.

Suppose the landlord has 300 acres worth $1500 per acre. A study of his buildings and livestock equipment shows them to be worth about $60,000 and they have a 10 year useful life. The tenant has a $120,000 machinery investment and figures a useful life—when it was purchased—of 8 years. Other figures are shown in the following table.

| | Contributions | |
	Landlord	Tenant
Land		
$450,000 × 5%	$22,500	—
Property tax	4,500	—
Business and equipment		
$60,000 × 10%	6,000	—
Labor and management	—	$18,000
Livestock		
$100,000 each × 10%	10,000	10,000
Machinery		
$120,000 ÷ 8	—	15,000
Totals	$43,000	$43,000

In this example, the calculations used show equal contribution by the landlord and tenant.

Naturally, not every arrangement is going to pencil out that easily. Then, you will want to look at a different than 50-50 split or adjust the contributions to make it a 50-50 deal.

Uneven shares can cause more difficulty in figuring each person's share of income and expenses. It's usually just easier to figure 50-50 on everything.

If the owner has a relatively small land base—say 120 acres of $1000 per acre land, his contribution at 5% is only $6000. If taxes are $1500 and all other figures are like in our example, his contribution is only $23,500.

If the tenant's contribution stays the same as in our example—$43,000—the tenant is providing nearly 65% and the owner 35%.

In that case, the owner might provide some more inputs such as labor, management, or facilities. Or he might provide more of the livestock to even out the input.

Another option is for the tenant to pay 65% of the other expenses—crop and livestock—and to take 65% of the income from both crops and livestock.

Bargaining

First thing is to make sure both landlord and tenant are satisfied with the input figures. The landlord may say the tenant's labor and management figure is too high. The tenant may counter that the land value or calculated interest return is too high.

Mutual agreement of what would be fair in a normal year should settle those problems. In a good year, each one will get more than those estimates. In a bad year, they may get less.

Fair market value of the land is often an arguing point. The figure you use for the caluclation should be realistic for farm use.

Say you use $3000 an acre at 5%—$150 per acre. Then you throw in $20 per acre for taxes. You might ask whether that land could really be cash rented for $170 per acre. Then the bargaining might result in a landlord contribution for land that's closer to the amount of cash rent he might reasonably expect if he were going to cash rent it.

As part of the bargaining process, be sure both parties understand these things:

(1) The crop program to be followed, including crop rotation and conservation practices.

(2) How records are to be kept and who is to keep them.

(3) The owner's amount of involvement in management. Does he want to materially participate for social security or estate planning reasons?

As you develop a lease and bargain on a final agreement,

keep some of the advantages and disadvantages of a livestock share lease in mind.

Advantages for the owner

His income can be increased with a livestock program if it is a successful one.

It also gives the owner another investment opportunity—an investment in livestock and facilities

A tenant who gets involved in a good livestock program is likely to stay longer.

Advantages for the tenant

Risk is less in the livestock program since he has a business associate who shares that risk.

The owner has his own money invested. Therefore, he's more likely to be interested in the livestock program and will be more willing to make improvements—in facilities, for example.

The owner may have livestock management experience to offer that will help make the program more successful.

It's a way for the tenant to get control of more capital for efficiency of volume in the livestock operation.

Disadvantages for the owner

The main obstacle for the owner is that he may have to sock more money into the operation. But that also gives him more potential for added income.

In some cases, he might find himself more actively participating in the farming than he wants to.

Disadvantages for the tenant

The tenant may feel his labor and management skills are giving the landlord too much return compared to his own return. The landlord's contribution in the form of buildings and equipment may not seem as important to him as his labor.

As income increases, the tenant may provide a bigger portion of the increased input than he receives in increased income. That may call for a reevaluation of the inputs and returns.

Disadvantages for both

As costs and size of the operation change, the arrangement may get out of proportion. One party may find he's providing 55% of the input for 50% of the return, for example.

The livestock share lease calls for a closer working relationship between the parties than with most leases. Therefore, the parties have to be more compatible to make it work. Both have to be able to "roll with the blows" on price and diseases, for example. They can't have an "in and out" complex.

Record keeping and financial settlement can be more complicated.

You have to have a solid agreement on who makes the final decisions if both parties are involved in management of the operation.

Summary

A livestock share lease should be tailored to fit the individual situation. It makes a big difference to the tenant if the land base is big or small relative to the livestock program. A small land base and large livestock program will tend to favor the landlord.

If the buildings and equipment are labor efficient, the tenant may be able to handle a good livestock volume without additional hired labor.

The kind of livestock will also influence the terms of the livestock share lease. With dairy, the tenant's labor input may be very high. On the other hand, a cattle feeding program with good facilities may be a low labor enterprise.

Consider, too, that the landlord may be in a stronger position to take risks than the tenant. He may want to feed more cattle or raise more hogs than the tenant can afford—or finance. That landlord may have to help finance the tenant in his share of the livestock program.

Finally, the fair livestock share lease is usually harder to develop than most other leases because of the many variations in different farming operations. But the effort can be profitable for both landlord and tenant.

Questions and Answers

Question: I own a beef cow herd and will furnish all feed for it. But my neighbor is going to provide all the labor. What would be a fair share of the income for him?

Answer: Labor costs typically average 10% to 15% of the total cost in a beef cow herd. That assumes average facilities and that the winter feed supply is at the feeding site. Therefore, a 15% share of the calf crop should be reasonable pay for the neighbor's labor.

Question: I own 200 cows in a 50-50 partnership with my son. He manages the herd, does all the work, and furnishes all the feed. What percent of the calves should I receive?

Answer: If you provide only 50% ownership of the cows, 10% to 15% of the total calf crop should be a fair return to you. If you are paying other costs such as veterinary expenses and providing buildings and feeding equipment, your contribution is more. Then, you would need to step up your percentage return based on the value of those contributions.

In other words, you should at least be reimbursed for any of the expenses you pay plus enough to at least cover your overhead costs on the buildings and equipment such as taxes, insurance, and repairs.

Question: The tenant on my farm takes care of 900 pigs farrowed each year and sold as 40 pound feeders. I own the hogs and pay all expenses. What is a fair percentage of the hog sales for each of us?

Answer: A share of 20% for the tenant and 80% for the owner would be about right. If the facilities for handling the pigs and feed are good, labor requirements will be about 1 to 1 1/2 hours per feeder pig raised.

That figures out to 900 to 1350 hours during the year.

If pigs sell for $40, for example, the tenant would get about $7200 a year, or a rate of $5 to $8 per hour.

Question: I have a new confinement building for 230 sows from gestation to 70 pounds. Hogs are then finished on concrete. The herdsman farrows 10 sows per week and provides all the labor. He owns one-fourth of all the stock and pays one-fourth of all the bills including feed, electricity, and veterinary fees. What would be a fair income split between the herdsman and myself?

Answer: Building and equipment costs may represent 10% of the total costs. Labor will add 15% to 18% of the total. Since all other expenses are on a 25-75 basis, a 1/3-2/3 split may be about right based on your contributions.

If the herdsman does an especially good job, you might increase his share to 35%. Or, you might take a look at some of the other incentive programs.

Question: How many 40 pound pigs per litter would be fair payment for someone to farrow my sows if that person furnishes feed, housing, electricity, water, and labor?

Answer: Do some careful figuring on this to calculate what percentage of the total costs each person provides. Chances are, you will find that 35% to 50% of the weaned pigs is about the right figure.

Another method is to keep track of the feed and pay that cost plus $1 per day per sow for labor and facilities. After you have done that for a year or two, you may be better able to determine a fair percentage of the weaned pigs.

Chapter 11

Labor And Profit Sharing Leases

Farmers and ranchers nearing retirement are constantly looking for a way to work a son, relative, or other young man into the farming operation. That's one place a labor or profit share type lease fits nicely into some farming arrangements.

You may view this as a father-son agreement rather than a lease, but it can also work with unrelated parties.

This type of agreement may serve as a ''testing'' stage—often for a father and son—to see if all parties are compatible. If it works out, you may want it to develop into a long term farming arrangement where you set up a formal partnership or even incorporate the operation.

In a typical labor share lease, the farm owner provides a fully equipped farm. The operator usually provides little or no capital to start with. So his portion of the return is usually based entirely on his labor and management input. The owner may still want to provide his own labor and management, too. Or, he may be ready to step aside or work only during the busy seasons.

The financial arrangement is the key to making a labor share arrangement work. If the owner's goal is to have the tenant take over eventually—maybe buy him out—his earnings are going to have to be good. Otherwise, he won't be able to build enough capital to make that longer range plan work.

Your compensation plan might be an income splitting plan. Or, it might be a salary and incentive or bonus plan. Let's look at some possibilities.

Splitting the returns between the owner and operator can be

figured two ways—one uses gross return, the other uses net return.

Gross return split

With the gross return method, the owner lists all his inputs—land, operating expense, labor and management, and equipment depreciation. By adding the value of the operator's labor and management, all the farming inputs are included.

The gross income is then divided in the same proportion as each person's contributions.

Here's a simplified example in which the owner has $500,000 worth of land. He plans to provide some labor. The partner (tenant) will work full time.

Contributions

	Owner	Partner
Land, $500,000 @ 5%	$25,000	——
Depreciation	10,000	——
Operating expenses (including property tax)	50,000	——
Labor @ $1000 per month	3,000	$12,000
Totals	$88,000	$12,000
% of total	88%	12%

In the example, the landowner provides 88% of the total inputs. The partner (tenant) furnishes the other 12% from his labor. Those percentages, then, are applied to the gross income to determine each partner's share.

Suppose the gross income is $120,000. Then the owner will get $105,600 and the tenant will get $14,400.

After the owner pays the $50,000 operating expense and covers $10,000 of depreciation, his return for investment in land and for his labor is $45,600. So he's getting his 5% return on the land plus much more than $1000 a month for his labor.

The input figures you use for contributions can make a big difference in that final percentage. For example, we used 12 months at $1000 per month for the tenant's labor input.

What happens if we change that to $1500 per month? The owner would contribute $89,500 (83%) and the tenant would contribute $18,000 (17%). If gross income is $120,000, the owner would get $99,600 of it and the tenant's income would be $20,400.

What this means is that you can build a lot of flexibility into this type of agreement. Depreciation will usually be actual figures from your tax return. Operating expenses will be estimated figures now, but actual figures at final settlement time. But the interest rate for return on your investment plus the labor and management rate per hour are extremely flexible. A low land value or interest rate or high labor and mangement rate will normally favor the tenant by increasing his percentage of the total calculated contributions.

If the partners are a father and son, the father may be willing to make that kind of a sacrifice—a deliberately "unfair" agreement. If it's a non-family member, you may have to do it, too, if you want him to farm with you. Otherwise, he may decide he'd be better off taking a job somewhere else.

Of course, be sure to consider "fringe" benefits he might get, too. If you provide him with a house and some meat for the freezer, for example, he should be satisfied with less actual cash. Besides, some of the fringe benefits aren't taxable.

Net return split

The second method of splitting income is based on the net return. It's similar except that a farm record or bank account is used. All farm operating expenses and depreciation are paid from this account. Then at settlement time between owner and operator, the net return is divided.

This payment is then based on the ratio of other contributions made. Looking at our earlier example, here's how the contributions would look.

Contributions

	Owner	Operator
Land, $500,000 @ 8%	$40,000	——
Labor @ $8 per hour	8,000	$20,000
Totals	$48,000	$20,000
% of total	70.6%	29.4%

The net return would be $80,000—as follows:

Gross income	$140,000
Expenses and depreciation	60,000
Income to share	$80,000

The owner's 70.6% would give him $56,480 to pay for his labor and management and return for investment. The operator would get $23,520 for his labor and management contributions.

Another net return method subtracts all expenses, including return on investment. In our example, that would leave $40,000 to be shared—gross income ($140,000) minus expenses ($50,000), depreciation ($10,000), and 8% return on investment ($40,000).

That remaining $40,000 is shared according to the proportion of the total labor furnished by each partner, or it could be arbitrarily split 50-50.

In this example, the owner provides 1000 hours of labor—two-sevenths of the total, or 28.6%. The tenant provides 2500 hours—the other five-sevenths, or 71.4%.

That means the owner gets $11,440 of the $40,000 profit ($40,000 × 28.6%). Plus, he is paid for all his expenses, depreciation, and receives an 8% return on his investment.

The operator gets $28,560 ($40,000 × 71.4%) for his labor and management contributions.

One other arrangement is to also take the labor out of the net, then share the remaining money equally.

In our example, the owner would get his $8000 for labor and management, the operator would get his $20,000. That would leave $12,000 to share equally. So each would get an additional $6000.

The owner would be repaid for his expenses, depreciation, and return on investment plus a total of $14,000 for his labor and management ($8000 wages plus $6000—half of the net income).

The operator would get his $20,000 wages, plus $6000 of the net income for a total of $26,000 for his labor and management contributions.

Inventory changes in an expanding operation should be considered on a regular basis, as they will affect the percentages of contribution. An annual review is probably best. If not annually, it should be done every 3 to 5 years.

Guidelines

The portion of payments the operator gets varies greatly from farm to farm. But a typical share of the gross return for the young operator runs 12% to 20%. If it's based on net return, the operator will usually get a 25% to 35% share of that net return.

An ambitious operator is going to want to grow into the operation even more. He's probably going to use some of his income as a down payment on machinery and land. The owner may even sell him some of his land.

As this happens, the operator's share of contributions increases. The owner's share decreases. This is normally consistent with the owner's plans to slow down and ease out of the operation.

By that time, the two business associates have probably also determined that they are compatible partners. So they may want to change to a more formal partnership or corporation arrangement.

Or the operator may have built up adequate assets and financial backing that make a crop share or cash rent lease more desirable. This can also be a good move when the owner decides he no longer wants to be involved in labor and management.

Other considerations

With high values or low-producing land, a high rate of interest return to the owner on the land can weigh the return heavily in favor of the owner. A too low wage rate can hurt the operator, too.

That means the owner has to look at his goals. If he has a strong desire to help a son or other young operator get started farming, he may be willing to figure a lower return on his investment. Or he might list the land at a lower than fair market value.

Good record keeping is an essential ingredient for a fair and trouble free labor share lease. That's especially true if shares are based on net return. Best bet is to set up a separate farm account and pass all farm income and expenses through it.

Consider what might happen in a loss year. Would both parties have a way to meet living expenses? Be sure to spell out exactly how gross or net income will be calculated—what expenses will be included.

Figures in our examples show only a total return for each party for the year. Most people can't live on one paycheck a year. So you'll likely need to set up a regular monthly wage for each partner—especially for the operator.

That wage might be figured as 50% to 75% of the expected total, for example. It should be calculated carefully to give the operator a liveable income throughout the year.

Same thing may be true for the owner. He can also be paid a wage for his labor and management so he can meet his personal obligations from current earnings.

Finally, keep flexibility in mind. Let your goals and objectives guide you in the figures you use for contributions. If you, as owner, want to give your operator—maybe your son—a fast start, especially after the first year or two, you can adjust the contribution figures to give him a bigger share. It's your farm, so you can do whatever you want to do with it.

Incentive and bonus arrangements

Maybe you don't like the bother or uncertainty of a share agreement at all. Then, you may want to look at a salary and bonus arrangement for that new partner (tenant).

Your objectives with a labor lease agreement aren't the same as they would be if you were simply trying to hire labor. You're getting a manager too. So you shouldn't think in terms of hired man wages.

This usually involves salary plus an incentive. The salary should probably be at least $900 to $1000 a month and maybe somewhat more, depending on the size and profitability of the operation. It has to be enough for the new person to meet his personal monthly expenses. Then you may want to add some type of an incentive program.

For example, suppose you set a solid salary base. Here are some guidelines for incentives you might want to pay in addition to salary. You may want to use one or more of these.

If the incentive is to be based on the whole farming operation, you might look at some of these bonus plans:

- 1/4% to 1% of the gross receipts as a bonus.
- 1% to 4% of the returns left after all cash operating expenses are deducted from gross income.
- 2% to 10% of the net farm income as computed on Schedule F of your tax return.
- A year end bonus of up to 10% of the cash wages the person was paid during the year.

Then his cash wages probably should be increased by up to 10% each year that he continues in the farming operation. Starting at $12,000 a year, annual 10% increases would put him at

$17,570 in his fifth year, for example.

If you want to base the incentive on a specific enterprise, there are several possible plans to follow.

On feeder pigs purchased and fed out, for example, the incentive bonus might be:
- 30¢ to 60¢ per feeder pig bought and fed out.
- 1/4% to 1% of the hog sales less the cost of feeder pigs.
- If death losses are less than 3%, the employee might receive 25% to 40% of the value of those market hogs saved. For example, if 98% of the feeder pigs purchased live to be sold on the market, the employee receives 25% to 40% of the market value of 1% of all the market hogs sold.

If yours is a feeder pig producing operation, the incentive might be:
- 50¢ to $1 per pig weaned.
- $2 to $3 per pig weaned above seven per litter.
- $4 to $6 per sow that weans more than eight pigs per litter.
- 1/4% to 1% of the gross income from hogs, including inventory changes.

For a complete hog program, the incentive bonus might be:
- 50¢ to 75¢ per hog marketed during the year.
- $3 to $6 per market hog sold above seven per litter.
- 1/4% to 1% of the gross income from hogs including inventory changes.

For a beef cow herd, possible incentives include:
- $2 to $6 for each beef calf weaned.
- 10% of the gross income above $160 per cow.
- $15 to $20 for each calf weaned over a 90% calf crop.

For feeder cattle purchased, the incentive might be:
- 25¢ to $1 per head of fed cattle marketed.
- 1/2% to 1 1/2% of the beef sales less cost of purchased feeders.
- 10% to 20% of the return over all cash costs including homegrown grain fed.

` Some common dairy incentives include:
- 1/2% to 2% of milk sales.
- 10¢ per hundred pounds of milk sold (this might be increased by 2¢ per year that the employee stays, up to a maximum of 50¢).
- 1% to 3% of the returns above feed costs based on DHIA records.
- 25¢ to 50¢ per hundred pounds of milk sold over 15,000 pounds per cow.
- $5 to $10 for each calf saved over a 90% calf crop at the end of the year.
- 15¢ to 25¢ for each 1000 pounds of milk sold over 300,000 pounds.

In your crop program, you might consider some of the following incentive bonuses:
- 1/2% to 1% of the grain produced, including grain equivalent in silage.
- 7¢ per bushel of corn produced over 100 bushels per acre.
- 12¢ per bushel of soybeans produced above 30 bushels per acre.
- $1 to $3 per ton of alfalfa that yields over 3.5 tons per acre.
- 1/8% to 1% of the gross crop sales.

These are just some guidelines for incentives. Some of these have been used with good success. You may want to develop new ones of your own or adjust these to fit your situation. Aim for incentives that will encourage better production—higher yields, more pigs per litter, reduced costs, etc. Then, both partners will benefit.

Finally, let's look at some of the advantages and disadvantages of labor share leases.

Advantages for the owner

Gradually working a new man into the operation lets the owner slowly back away from the labor and management chores if he wants to. Or, if he wants to continue at full speed, he has the advantage of a partner who also provides management rather than just hired labor.

It gives the owner a way to keep the farm operation he has worked hard to build together in a solid unit.

The owner can keep close supervision during the changeover period. At the same time, he's training his replacement.

Biggest use of the labor share agreement is to shift management—and maybe farm property—to a son or a daughter.

There can be a great deal of satisfaction in helping a young person work into an operation you have spent years developing—and comfort in knowing it will continue.

Advantages for the operator

With the high cost of equipment and land, a labor share lease allows the hard working manager a chance to develop his farming skills, earn a living, and build up his capital simultaneously.

He gets the benefit of supervision by a successful operator.

Then, as his farming expertise develops and his contributions in labor and equipment increase, his share of the profit pie gets bigger.

He can get started farming with a minimum of risk and a minimum of capital investment. In fact, it may be the only way he can get started farming on a solid footing.

Disadvantages for the owner

The most common problem for the owner is that an operator who is short on capital is often shy on experience, too. This means he may have to provide more management than he would like in the early years. Then, if the owner has to materially participate, he could lose some of his social security benefits if he's of retirement age.

It takes careful planning to assure that the agreement is workable and satisfactory to both parties.

There is a definite need for farm records. This takes a lot of time and trust by both parties involved.

Before committing yourself to such an arrangement, you should make sure the income possibilities are big enough to support both families.

Disadvantages for the operator

The operator may be anxious for more responsibility and pay. The owner, however, may not be able or willing to provide this.

He may feel too limited on decision making in the early years while the owner still wants to be actively involved.

The owner and operator may have conflicting goals concerning new technology and expansion.

Summary

It can take a lot of figuring to develop a fair labor share lease arrangement for both landlord and tenant, but it can be flexible. And the final results can meet some other important family and farming goals.

Chapter 12

Pasture Rent

Pasture grass isn't as easy to price as corn, wheat, or bales of hay. But it does have a value. And there are some good guidelines to help you figure a fair pasture rent.

The three basic ways to figure pasture rents are:

(1) Per head of livestock per month.

(2) A flat rental rate per acre.

(3) By weight gain of the livestock.

Let's look at each of the methods and some examples.

Per head of livestock

A University of Nebraska formula is probably the most useful and accurate for figuring a fair charge per head of livestock per month. It considers the size of animal and the quality of the pasture.

The formula looks like this: Average weight in thousands of pounds during the pasture season × average price of good hay per ton during the pasture season × pasture quality factor = the rate per head per month.

The pasture quality factors are:

- Lush, green, high protein pasture, 0.22.
- Excellent tall grass pasture, 0.20.
- Fair to good native pasture, mostly short grass, 0.15.
- Poor short grasses or considerable weed growth, 0.12.

For example, suppose you're figuring cattle that will average 900 pounds during the pasture season. Good hay averages

$50 a ton. The pasture quality is fair to good predominantly short grass pasture (a 0.15 rating). The formula will give this result: (0.9 × $50 × 0.15 = $8.10).

So a fair rental would be about $8.10 per head per month.

Another way

Another per head per month rent method sets the rate for an "animal unit" and then adjusts for size of animal. A 1000 pound animal is considered to be one "animal unit."

First, we figure a fair rental rate for that animal unit. Here are some rule of thumb guidelines.

- Rental rate per month is 2.2 × the price of a bushel of corn.
- Rental rate per month is the price of a ton of hay ÷ 8.5.
- Rental rate per month is the price of fed cattle ÷ by 11.

If corn is $2.50 per bushel, the rate for one animal unit would be $5.50 per head per month ($2.50 × 2.2). If a ton of hay is worth $50, that rate would be $5.88 ($50 ÷ 8.5). If fed cattle are selling for $60 per 100 pounds, the rate would be $5.45 ($60 ÷ 11).

Once you have that rate for a 1000 pound animal unit, you can use the following table to adjust for other size animals.

Age	Weight	Mature animal unit eqivalents
Mature bull	1400-1600	1.4
Mature cow	1200	1.2
Mature cow	1000	1.0
Mature cow	900	0.9
Heifers (1-2)	700	0.7
Heifer under 1 year	600	0.6
Calve	400	0.35
Calve	300	0.25

For example, say the animal unit rate is set at $5.50. But, if 700 pound heifers were being pastured, a fair rental for them would be $3.85 per head per month ($5.50 × 0.7).

This table can also help you determine the stocking rate for the pasture. For example, suppose you figure an acre of your pasture will handle one mature 1000 pound cow. Then one acre should handle about 1.4 heifer units weighing 700 pounds (1 ÷ 0.7) or 0.8 mature cows (1 ÷ 1.2). Looked at another way, it's going to take 1.2 acres for each mature 1200 pound cow.

Rent per acre

Renting pasture by the acre may take less figuring. But it may tempt the renter to overgraze the pasture to get his money's worth. After all, whether he runs one or three head per acre, it makes no difference in his rent.

Obviously, the landlord doesn't want the pasture to be over-stocked. It hurts long term production. But, since rainfall has such an important impact in pasture production, it's hard to know how many head a pasture will actually support. To protect his pasture, the owner may want to set a maximum number of animals that can be grazed, or starting and stopping dates for grazing. That limit should be based on the pasture quality and size of animals.

But how much do you charge per acre? Economists and farm managers say the annual rental rate per acre for pasture will average between 4% and 6% of the sale value of the land.

For example, pastureland that would sell for $600 per acre should draw $24 to $36 per acre cash rent for the grazing season.

That's a good starting point for average pasture. If it's improved by fertilizer, chemical weed control, or new seeding of higher producing grass, the landowner needs to at least recover those costs, too.

Another method to determine per acre rent is to add up the landowner's costs for taxes, insurance, improvements, and upkeep. Then you'll need to add at least 2% to 4% of the land value for return on his investment.

Weight gain method

The weight gain method bases pasture rent on the actual weight gain of the livestock. Obviously, this is suitable only for feeder cattle, not cows.

The cost of gain per pound needs to be negotiated. And it's necessary for the cattle to be weighed on and off the pasture to determine payment. That may not be practical for many farmers.

A good estimate for a fair rent is 15¢ to 17¢ per pound of gain.

For example, say the average gain is 1.25 pounds per day. At the 15¢ rate, that would be about 18.75¢ per day or $5.62 per head per month.

One problem is that the renter might put a bunch of slow gaining cattle on the pasture. That could shortchange the landowner.

Also ask: What about animals that die? If a steer has gained 100 pounds, the landowner should get $15. But that weight won't go over the scale with the other animals at the end of the pasture season.

Who does what?

The responsibilities of both the owner and renter are usually negotiable.

Generally, the landlord provides water, materials for fence repairs, and other facilities on the land.

The tenant provides the labor to maintain fences if the fences are in good shape to start with. He also supplies all the salt and minerals used by the livestock. The tenant should also handle insect and weed control—clipping when needed.

Be sure to stipulate who is to check the cattle, water supply, and fences. And who is to call the veterinarian if needed? Usually, that's the tenant's job. But if the landlord does it, he needs to receive a fair payment for his time.

Problems to avoid

Even after the price terms are worked out, there can be some areas of dispute in pasture rental. One of these can be pasture maintenance.

The lease should define who is responsible for mowing the weeds and controlling brush. The contract should also spell out the terms of fertilization, including the rates. You may even want to get a soil test before signing the lease. That way, both parties know what fertility is needed.

The landowner may want to write in special provisions specifying when he can have cattle taken off the rented pasture if it's under stress from weather or if it's getting late in the season. Sometimes an owner will want a clause in the contract specifying how short a pasture can be grazed.

Of course, local supply of and demand for pasture to rent will have a strong influence on pasture rent. Pasture grass usually can't be stored for later use. So negotiation can become a big part of setting the final rent figure.

If there's only one person who wants to rent the pasture, he may shoot for a real bargain. But that could come back to haunt him the next year if the landlord finds someone else to rent that pasture. He may not even give the previous renter a chance if he felt shortchanged by him the year before.

Finally, like any other lease agreement, pasture rent should be covered by a written lease. There are a lot of details—grazing restrictions, weed control, death losses, and who provides what. It's too easy for one party or the other to forget what the original agreement was if all you have is an oral lease.

Questions and Answers

Question: I rent some pasture that is improved well above average to a dairyman with high producing cows. What is a fair rent per month by the head?

Answer: Good pasture that is adequate for high producing dairy cows should be worth $7.50 to $10 per head per month.

Question: My neighbor wants to rent 35 acres of corn stalk ground this winter as pasture for his cattle. How much should I ask as rent?

Answer: Quality of stalks left as feed, amount of corn left, snowcover, fences, and availability of water are all important considerations.

If you can meet all those things, a price of $3 to $5 per cow per month would still be cheaper than hay for the owner. A straight rent per acre might be $8 to $12 for the season.

If the field isn't available until late in the season, the price will be lower. You may also want to restrict use of the field to a March 1 cutoff since you don't want cattle in wet fields after the spring thaw.

Question: What would be a fair price per acre for a good field of alfalfa and orchardgrass rented out for pasture?

Answer: If it is good productive land and the alfalfa mix is safe to pasture, it might carry one animal unit per acre for 6 or 7 months. Then it should be worth $40 to $50 per acre for the pasture season.

Chapter 13

Building Rent

Farm owners often ask: "How much rent should I charge for a farm building—or for a set of buildings?"

It is tough to figure what buildings are worth. The income they will generate isn't as predictable as for land. Plus, there's usually a limited demand for farm buildings. Maybe only one person is interested in a particular building or set of buildings.

Result is a rather wide range in building rent—from what it's really worth to the tenant to what the owner can get. But there are guidelines to help you set a price that's reasonable to both the owner and the tenant.

There are two key things to consider:

(1) Building ownership costs.

(2) How much the building or buildings will contribute to farm returns.

Annual ownership costs are fairly easy to figure. They're made up of two parts—fixed and variable costs.

Fixed costs

Fixed costs continue whether or not the building is used. Think of them as the DIRTI 5: Depreciation, Interest, Repairs, Taxes, and Insurance.

Depreciation is used to measure deterioration. On new buildings, the annual depreciation would be the cost of the building divided by its useful life, if you use straight line depreciation. Suppose the building cost $10,000 and you figure its useful life is 25 years. Annual depreciation charge would be $400 ($10,000 ÷ 25). You can convert the depreciation to a percentage by dividing 100% by the useful life. In this example, depreciation would be 4% a year (100% ÷ 25 years). Then apply that to the $10,000 cost to get your $400 a year depreciation. With older buildings, divide the present value of the building by the remaining years of life. A building worth $3000 that can still be used for 10 years, for example, would have an annual depreciation cost of $300.

Interest is the going rate of interest times the average investment in the building over its expected useful life. For easy figuring, you can take the going interest rate times half the original cost of the building. For example, suppose the original cost was $10,000 and you use a 9% interest rate. Interest might be figured as $450 a year ($5000 × 9%).

Repairs will have to be made to maintain the building in useable condition even when the building is not being used. A rate of 1 1/2% to 2% is probably going to be in the ballpark. The rate on confinement livestock buildings may be higher than that. On grain storage, it may be lower.

Taxes are figured by applying your local property tax to the assessed value (not necessarily the actual value) of the building. A 1% original cost or present value is an often used rate for estimates.

Insurance costs can be obtained from your insurance policy. You can figure that insurance cost will probably run in the neighborhood of 1/2% to 1% of the value of the building.

The only way to stop these annual fixed costs is to tear the building down. Trouble is, that cost may exceed the salvage value of the structure.

As an example, let's consider a set of modern, specialized swine buildings. By farrowing quarterly, the buildings are capable of handling a 50-sow unit. The initial cost of the buildings three years ago was $50,000. The owner's estimated annual costs are shown in the table below.

Depreciation (15 year life)	6.7% x $50,000 =	$3,350
Interest (9% rate)	9.0% x $25,000 =	2,250
Repairs	1.5% x $50,000 =	750
Taxes	1.0% x $50,000 =	500
Insurance	0.5% x $50,000 =	250
Total	14.2%	= $7,100

Actual out of pocket cash costs—to cover repairs, taxes, and insurance—are $1500 in this example. So the owner might have an asking price of $7100 for rent. But his bargaining range might go as low as $1500.

If a building contains equipment, you may want to break it out and calculate the DIRTI 5 separately. Equipment wears out faster than the building itself so it should be depreciated faster. Repair costs are also likely to be higher. Therefore, a different set of percentages should be applied to the equipment than to the building.

Let's say the building described is worth $40,000 and has $10,000 worth of equipment in it. Say the equipment has a 10 year expected life. These percentage of cost figures will be more accurate.

	Building	Equipment
Depreciation	6.7%	10.0%
Interest	4.5	4.5
Repairs	1.5	3.0
Taxes	1.0	1.0
Insurance	0.5	0.5
Totals	14.2%	19.0%

When you apply 14.2% to the $40,000 building value, you get a cost of $5680. Add 19% of the $10,000 worth of equipment ($1900) and you get a total of $7580 as the top asking price for rent. Actual out of pocket costs are estimated at $1650 ($40,000 × 3% plus $10,000 × 4.5%). The owner would need $1650 to cover repairs, taxes, and insurance.

The percentage figures used here are guides. Your own ownership records will indicate actual costs. Repairs, taxes, and insurance can be found in your cash expenditures. Depreciation should be listed on your depreciation schedule for tax purposes.

What value to use

On older buildings, it can be tough to figure a fair value to use. Should you use the undepreciated value? Or, should you use something else?

Undepreciated value is probably the easiest to find. You can go to your depreciation schedule. Original cost minus total depreciation taken over the years is the undepreciated value. Trouble is, that undepreciated value may be a lot lower than what the building is actually worth. In fact, you may have even depreciated it out so it shows little or no value.

Then you may want to look at its actual fair market value. An appraiser could give you the answer. The appraised value for property tax may be another guideline.

Another way is to look at its replacement cost. What would it cost today to build a new one like it? Then depreciate that. Say it would cost $18,000 to replace it, your building is now 20 years old, and looks like it will last another 10 years. You have one-third of its usefulness left, so you might figure it's worth about $6000 ($18,000 replacement cost × 1/3).

If you figured a 17.5% rate on that, rent would be $1050 a year. That's 10% for depreciation, 9% on one-half the value for interest on investment, 1.5% for repairs, 1% for taxes, and 0.5% for insurance.

Variable costs

Variable costs are those that change with the use of the building. For the most part, they are added repairs caused by use. But they might also include use of electricity and water. Rental you ask for a building should at least cover these direct cash costs. Or the renter might pay those extra variable cash costs.

If you're the renter

On the renter side of a building lease, you have to ask other questions: How much can you afford to pay? How much do you have to pay?

What you can afford to pay depends on the value of the building to you.

You may face two possible situations. First, you have a need for a building, but don't have one. Or maybe you have a building you are presently using but it doesn't fully fit your needs. Maybe the location is bad. Or maybe there isn't a reliable water supply. So you would rent a better building if it were available at the right price.

Consider having versus not having a building. If, after all the costs are considered, you figure you could make some money or save some money by having a building, that's the maximum amount you can afford to offer. Say you figure you could save 20¢ a bushel by renting grain storage from a neighbor rather than storing the grain in town. Your savings might include storage cost, time at harvest, and transportation. That 20¢ per bushel is the top limit you could afford to pay for the neighbor's storage.

Another way to figure rental value of a building is to estimate how much income it will produce.

Let's go back to our earlier example of a modern, specialized swine facility. Based on a projected hog market (3 to 5 year average) and feed costs of $22 per hundredweight, here are some rough estimates of how much the renter can afford to pay.

Gross Sales		$63,000
Feed costs	$36,400	
Other cash costs	8,000	
Labor	5,200	
Management	3,000	
Interest on operating expenses	1,600	
Interest on sow herd	500	
Fixed costs (machinery & equipment	2,000	
Total Costs	$56,700	
Dollars available for building rent		$6,300

As you can see, that's less than the $7100 owner's cost shown earlier. So the landlord and tenant may be in for some give and take bargaining.

But there's a good amount of room for bargaining in this example. Based on our figures, here's the bargaining range for both parties.

Owner	$1,500 to $7,100
Renter	0 to $6,300

For the renter, any amount over $6300 will cut into his labor and management return. So a fair rent would cover at least the cash ownership costs of the buildings ($1500 in this example). But it needs to be less than the owner's total annual ownership costs ($7100) if the renter is going to make money with it.

Of course, you'll want to use figures from your own records. Then, both owner and renter should study them to be satisfied that the income, expense, and labor and management figures are reasonable. For example, the building owner probably isn't going to be happy with the renter's "top rent" figure if he has calculated a labor and management return at $25 an hour.

Grain storage

Rent for ear corn storage space will normally range from 6¢ to 10¢ a bushel for the year. On a per month basis, you might step that up 1 1/2¢ per month, figuring it may be used only a few months.

For shelled corn or other small grain storage, 12¢ to 18¢ per bushel a year has been a common figure. On a monthly basis, figure 1¢ to 2¢ per bushel minimum storage charge.

If drying, loading, and unloading equipment is provided, best bet is to figure an additional charge based on ownership costs in the same manner as we calculated for equipment in a building.

For example, suppose there's $8000 worth of that equipment. Say ownership costs for depreciation, interest, repairs, taxes, and insurance figure out to 19%. A $1520 charge for use of that equipment—above the storage cost—might be reasonable.

Normally, the renter will also pay the cost of electricity for drying and handling the grain in and out of storage. If the owner pays for it, he should be reimbursed.

Silo rental

A rent of $2 to $3 a ton for one filling a year is normally considered reasonable for a silo. If the silo is filled more than once a year, a higher rate might be charged.

Another alternative is to figure annual ownership costs the same as was figured earlier for the hog facility.

Machine shed

A typical rental rate for machinery storage is 12¢ to 15¢ a square foot for a open front, dirt floor building. An enclosed building with a concrete floor could demand a 20¢ per square foot rate.

You may also want to calculate ownership costs to make sure the square foot figure is in the ballpark.

Other considerations

Demand and location are two other important things to consider in setting the annual rental rate on any building.

If only one person is interested in renting a building, there's no competition factor—and maybe no good alternative for the owner. That may pull the rental rate far below these guidelines.

In that case, the owner should at least try to cover his out of pocket costs. But he may have to settle for a lot less than the total annual cost of the building.

Location of the building can make a lot of difference. If it's just right for the renter, he can probably rent cheaper than he can build a similar structure. But if you, as the renter, have to go out of your way to be able to use it, consider the inconvenience costs such as time and travel. Balance that against the cost of a new building when you bargain on rental rate.

Buildings with land

Often, a whole farm is rented for cash or on a crop share basis with buildings included.

Cash rent for the value of the buildings may be reflected in a higher land rent than if the farm didn't have buildings. Or the owner and tenant might set cash rent rate for the land plus an added cash rent for all the buildings.

If the renter doesn't need or want to use the buildings, the owner might rent them to someone else.

It used to be a common arrangement for the tenant in a crop share lease to get use of the buildings at no extra cost. While it is less common now, it's still a practice in many cases. Then grain storage facilities are shared by the owner and renter in the same proportion the crop and expenses are shared.

But, if there's a good house and/or livestock facilities that the tenant uses—with no benefit for the owner—that owner is justified in charging some additional cash rent for the buildings. That may depend on the quality of the land, too. With lower quality land, the tenant may pay a low rent or no rent at all on the buildings to compensate for the lower quality land. That's especially true if the buildings are just average.

Always agree on who is to provide the water supply and keep up the water system.

Complete set of buildings

Another possiblity is to have land rented to one party and the buildings rented to someone else. Maybe the person who rents the buildings even lives in the house. Maybe the person who rents the land simply isn't interested in the buildings.

Basically, the same rules for a fair rent apply as if you're renting just one building. But now you estimate the owner's actual out of pocket costs for the total set of buildings. You also figure his potential interest and depreciation return. That gives you his bargaining range.

The renter can also figure his potential returns as a result of using the buildings—as we demonstrated on the hog facility. That gives him an estimate of his bargaining range.

From there it becomes a bargaining situation. The total rent should fall somewhere within the range of the owner's out of pocket costs and the renter's estimated returns plus a fair rent for the house if it's included.

House rent

The rental rate for a dwelling is usually easier to determine. The going rate in your community is the best guideline. There's also a good demand for country living from people who work in the city.

Bargaining

You can probably see that there's no totally accurate way to set a rent figure for farm buildings. But the guidelines in this chapter should be helpful.

Neither party should go to extremes in seeking a good deal. That can cause hard feelings that might haunt you later. If the owner feels shortchanged, he will be reluctant to spend money on repairs and improvements. Plus, he may quickly rent to someone else and kick the present renter off if a better deal comes along.

Same thing works in reverse for the renter. If the rent seems too high, he may overuse the facilities and cause more damage than usual. And he may keep looking for a better deal.

Good bargaining that results in a rent figure that seems fair to both owner and renter can avoid those kind of problems.

As a last resort, you may even want to seek some outside help to establish a fair and reasonable rent. A good way is for the owner to choose one person and the renter to choose one person. Then those two people can choose a third person. That will give you an impartial team.

Size of the farm makes a difference, too. Building rent is more critical when there is a good set of buildings with a fairly small land base. Typical land rent may not be enough to compensate the landlord adequately for his investment in buildings.

With a big land base, the landlord may be willing to take a low building rent as a way to attract a top tenant.

Finally, determine in your lease whether or not the tenant is allowed to sublease the buildings, especially if a good house is involved. The agreement might state that the landlord has to give written permission before his tenant can sublease any buildings.

Questions and Answers

Question: I have two farms rented out on a 50-50 basis. The tenant has full use of all buildings on both farms except the home rented out separately on one. The tenant has complete control of his hog operation plus the use of 30 acres of good pasture. I don't share in the receipts of any of the livestock. Should I receive rent for use of buildings and pasture used in connection with the hog operation?

Answer: There's no set pattern for rent of buildings on farms rented on a 50-50 crop share basis. In some cases, landlords figure their 50% share of the crop compensates them for the use of the buildings. In others, there is an additional rental rate. Therefore, look at the total rent package to see if you feel that it is fair or not. If you think you're getting the short end, bargain with your tenant for additional rent on the buildings.

You should definitely have cash rent for land in pasture and farmstead that the tenant is using for his livestock program. The chapter on pasture rent should be helpful.

Finally, consider the buildings. If they are in good shape and you pay the taxes, insurance, and upkeep costs, some rent could be justified.

Question: I have a corn crib and grain storage building that will hold 6000 bushels of ear corn and 3000 bushels of shelled corn. If I furnish electricity for loading and unloading grain, what should I charge per bushel for storage?

Answer: On an annual basis, you should charge 12¢ to 15¢ per bushel. Consider the demand for storage in your area. You may be able to charge a higher rate for shelled corn storage if corn is put in a grain reserve program that ties up your storage for a longer period.

Another way to charge might be 1¢ to 1 1/2¢ per month with a minimum of 6¢ per bushel.

Question: My landlord and I would like to put up drying and storage facilities for 20,000 bushels of corn. Should the landlord furnish the facilities and charge me for their use? Or should the landlord furnish storage and the tenant furnish the dryer? And if we go that route, what should each charge the other for the use of the equipment?

Answer: Drying costs should be shared. If the dryer is part of a permanent installation, it's usually better if it belongs to the landlord. The tenant should then pay half of the power and fuel costs and all maintenance costs. If the dryer is portable, then it would be jointly owned and all costs shared equally.

If the landlord furnishes permanent storage, he might charge 12¢ to 15¢ per bushel per year for storing the tenant's grain. The tenant might then furnish the dryer and fuel and charge 1 1/2¢ per point of moisture removed.

Question: I have 160 acres with a 24 unit farrowing house with two pits, a six pen nursery, a semi-confinement finishing house, a 32 x 60 foot cattleshed, an 8000 bushel drying bin, a modern house, and several other good buildings. What would be a fair rent with land on a crop share basis?

Answer: Buildings other than grain storage and obsolete facilities should rent for roughly 7% of their value. If they are insured for $50,000, about $3500 a year would be reasonable if a specialized hog producer can be found in your situation.

Often the landlord would not receive rent for grain storage and probably for machine sheds on a crop share rented farm. In those cases, the grain storage is usually shared equally by the landlord and tenant.

Another way to look at some of the facilities might be to charge about $75 per litter of capacity for the swine facilities. Then other structures could be rented on the basis of their value.

Question: A neighbor stores bales of hay in our barn, getting it as he needs it. What rent per bale is reasonable for this storage? He also messes up our yard somewhat, plus there's wear on our hayloft. Who pays for these repairs?

Answer: A rent of $5 to $7 per ton or 12¢ to 17¢ per 50 pound bale should be about right. If repairs are the result of normal use, they would be your responsibility. If caused by carelessness on the part of the neighbor, then he should be liable for these repairs. He should also take care of the yard damage.

Question: I have a large modern stanchion barn with a milk house. My neighbor would like to use these dairy facilities. What would be a fair charge?

Answer: Building rent of $75 to $125 per dairy cow unit per year would not be out of line. Otherwise, you might look at a rate of about 10% of the present value of the facilities.

Question: My tenant wants to rent my 18 x 55 foot cement stave silo. It is equipped with a silage unloader. What is a fair price?

Answer: That silo has a capacity of about 300 tons of corn silage. A rental charge of between $2 and $3 per ton for one filling a year would be reasonable. If filled more than once a year, a higher rate would be justified for the extra wear and tear on the structure and equipment.

Chapter 14

Building Up A Rented Farm

"The farm I'm renting needs work. Fertility is low. Wet spots take extra time and cut yields. And, I'd sure like some on-farm grain storage. Trouble is, my landlord doesn't want to spend the money."

That's a common dilemma. It takes some real landlord-tenant cooperation to work it out. But it can be done—usually to the benefit of both.

Requests from tenants to landlords for improvements typically include projects that will expand the operation, produce better yields, cut costs, or save labor.

Most questions regarding improvements on a rented farm fall into three areas:

(1) Building up fertility.

(2) Land improvments such as tile, terraces, and other conservation practices.

(3) New buildings or major remodeling of old buildings.

Let's take a look at each of these separately.

Fertility

Maybe a retiring farmer or previous tenant has squeezed the last bit of fertilizer out before leaving the farm. It's now going to take more than the normal amount of fertilizer, and possibly lime, to rebuild the fertility of that soil.

If the landlord is willing, you might agree that the landlord pay the cost of building the fertility up to average levels, then split the cost of additional fertilizer as you normally would in a 50-50 or other crop share lease. If the tenant is cash renting, the landlord might build the fertility level up to that average level, then the tenant would pay for the additional fertilizer he wants to apply.

But the landlord might not like that idea.

Then the tenant might pay his share for building the soil up to productive levels. His share would depend on the kind of lease you have. In a 50-50 crop share lease, for example, the tenant and landlord would share the buildup expense equally. In a cash lease, the tenant would bear the whole cost.

Trouble is, it may take several years for the tenant to recapture his extra cost through higher yields. He needs some time protection to be sure he will recover his cost. A 3 year or longer lease agreement might assure him he's going to be able to farm that land long enough to cover most or all of his additional cost. Or the landlord might agree to a lower rent for 2 or 3 years while the soil fertility is being restored.

Another possible solution is to make an agreement that the tenant will be reimbursed by the landlord for 30% to 40% of the amount he spent for phosphorus and potash applied the previous year when the lease is terminated. That's roughly how much of those two nutrients will be carried over to the following year. If he leaves after the first year, he gets a good share of that buildup cost back.

Nitrogen has less carryover value. But since the tenant may have spent extra money just to build the nitrogen level up to normal conditions, the landlord might reimburse him for one-fourth to one-third the amount it took to build up to that normal level.

Any reimbursement agreement should be put in writing before the lease starts.

In rental situations where fertility is at normal levels when a new tenant arrives, the tenant shouldn't expect any reimbursement for fertility when he leaves the farm. He's simply expected to leave the land as productive as it was when he started farming it.

If you're the landlord, consider that getting fertility built up to a good level is going to increase your return for many years in a crop share agreement. If you are cash renting the land, you'll attract better tenants and be able to get a more attractive cash rent if the farm carries a good fertility level.

Both landlord and tenant should carefully consider the value of the right fertility level. If an extra $10 spent for fertilizer will return $20 worth of extra yield, it's certainly a good investment.

Land improvements

Soil improvements such as tile, terraces, and water diversion channels can increase both yields and farmable acres. Therefore, both the landlord and tenant should benefit from these improvements.

That raises the big question of who should pay for those improvements.

Typically, the landlord pays for land improvements. However, a landlord is sometimes reluctant to make an investment of that size unless he can see enough additional return to pay for it over a fairly short period of time. If he's in retirement, that may eat up a sizeable chunk of his available cash.

However, that kind of improvement usually increases the long term value of the farm. In fact, failure to make these improvements can reduce the land value because of erosion or simply general deterioration. So the landlord should usually pay the cost.

If he simply can't, one answer might be for the tenant to pay the cost or part of the cost for some of the improvements that increase production. Then he should have a written agreement with the landlord that he will be reimbursed so much per year that he farms the land. This might be based on the increased yields the landlord will get from the improvement.

For example, suppose tiling costs $10,000 and increases the total farm yield by 1000 bushels of corn a year. In a crop share agreement, the landlord would get 500 bushels of that. He might agree to reimburse the tenant $1000 per year until his investment is paid.

If the tenant leaves the farm after three years, the agreement might call for the landlord to pay him the remaining cost in one lump sum.

In a few cases, the tenant might agree to pay the cost in return for a long term lease, say 5 or 10 years. That way, he can expect to get his money back in terms of increased yield, reduced cost, less wear and tear on his machinery, possible additional acres as a result of the improvement, and reduced labor. Pencil it out to estimate the number of years it will take to recover the cost.

In a cash rent situation, the tenant might agree to pay an extra $5 to $10 per acre rent if his landlord makes some of those improvements.

There are a number of variations you might use. It becomes a matter of bargaining until both parties are satisfied.

The landlord should also consider that making these land improvements may help him keep a good tenant on the farm. The tenant won't be as likely to move to a better farm if one becomes available.

Again, make sure all reimbursement agreements are part of the written lease.

Buildings

In both increased fertility and land improvements, the landlord can expect to share in the increased returns or reduced costs of those improvements in terms of a higher cash rent per acre, better quality tenants, and increased value of his land.

With building improvements, the benefits may be more one sided in favor of the tenant.

However, grain storage facilities can have benefits for the landlord, too. And in a livestock share lease, additional buildings should increase the landlord's return.

But in many building situations, the pendulum may swing more toward the tenant either paying for the structure or agreeing to some added rent to reimburse the landlord for his cost.

Most leases specify that the landlord is to approve permanent improvements even if the tenant agrees to pay for them. This would be true for land improvements as well as for buildings.

But beyond that, provisions should be made in the lease for some kind of settlement between landlord and tenant in case the lease is terminated by either of them. The forms in the back of this book include a lease supplement for making improvements on a rented farm.

For property that can be moved off the farm if the tenant leaves, the tenant may agree to pay the cost. Then he takes the structure with him when he leaves.

For example, suppose a tenant wants some grain storage bins. He might pay the cost of putting them on the landlord's property, with the stipulation that if he leaves, he can dismantle the bins and take them with him.

This might also make the tenant's banker more agreeable to financing improvements on someone else's property.

There also needs to be protection for both landlord and tenant. For example, the lease should specify a time limit for removing the property after the lease expires. This should be written to give the tenant adequate time in case of bad weather or other complications that delay moving the property. But it shouldn't be too long since the new tenant may want to use that same space for something else.

There should also be an agreement that the tenant is to leave the landlord's property in the same condition as it was in when the tenant rented the property. In other words, the tenant should repair any damage done in the removal process. If removal leaves a mess, the tenant cleans it up.

Make sure all these things are clearly spelled out in writing. Both the landlord and tenant should agree to them before any improvements are made.

With buildings, a tenant's biggest concern about paying for improvements on a landlord's property is that his lease might be terminated before he can get full value out of that improvement. He naturally wants a fair return for the money and time he invested in it.

Reimbursement plan

The improvement may be so permanent on the land that there's no way the tenant can take it with him when he leaves. Here's where a reimbursement plan might work.

For example, the tenant might pay the cost of the improvement. Then the landlord agrees to reimburse the tenant for part of the cost if he leaves the farm before he can recover his investment.

Here's an example how that might work: Suppose the tenant wants a new hog facility and spends $20,000 to construct it. If the tenant leaves after 6 years, they figure he has received 60% of the value of the building, or $12,000 in this example ($20,000 × 60%). So the landlord pays the tenant the remaining $8000 of undepreciated value.

A slight variation of this might be to use a different depreciation method, such as a declining balance method, that writes off more of the cost in the earlier years. This could be fair in the case of buildings that might become obsolete. But the fast write off on most buildings tends to be unfair to the tenant.

Here's an example using double declining balance method depreciation. You simply take the straight line method of depreciation and figure the percentage. In this case you would divide 100 by the 10 year useful life. That gives you a 10% straight line depreciation each year. Double declining balance simply doubles that percentage figure and applies that to the undepreciated value each year.

End of year	Undepreciated value	Depreciation
1	$10,000 x 20%	$2000
2	8,000 x 20%	1600
3	6,400 x 20%	1280
4	5,120 x 20%	1024
5	4,096 x 20%	819
6	3,277 x 20%	655
7	2,622 x 20%	524
8	2,098 x 20%	420
9	1,678 x 20%	336
10	1,342 x 20%	268
11	1,074	

If the tenant left after 6 years, for example, the landlord might pay him the undepreciated value of $2622. Under the straight line method, he would pay $4000.

With less specialized buildings, the depreciation rate might be somewhat less. It might be a building that will last for 20 years or more and be useful to almost any tenant who might later rent the land.

Now let's look at the landlord's situation with that $4000—or $2622 with declining balance depreciation—that he pays the previous tenant. How does he get it back?

He could be stuck with that as an expense if he can't find a tenant who wants to use those facilities. He has two options if the new tenant wants to use the facility.

One choice is to offer the tenant a chance to buy the outgoing tenant's interest. A second choice is to simply charge the new tenant additional rent to cover that investment.

The landlord runs a risk of losing if the new tenant isn't able to or doesn't want to use that building. Therefore, the lease with the original tenant who constructed the building might specify a penalty if the tenant is the one who terminates the lease. In other words, the lease might specify a 25% to 50% lower payoff by the landlord unless the landlord can find a new tenant who will pay extra for use of the facilities.

In our example, he might only pay the tenant who leaves $3000 for his remaining value in that hog facility rather than the $4000—if straight line depreciation is used for the reimbursement plan.

In any situation, you have to look at who gets the benefits from the facility. If the landlord gets some of the benefit over the years, he might agree to pay a percentage of the original cost at the time the facility is constructed. That percentage should be based on the share of the benefits from it that he will get. Or, if the tenant pays the cost, he might get a bigger percentage—or all—of the increased income from the improvement.

In a livestock share lease where the landlord normally provides facilities, most added facilities would be the landlord's expense.

One other alternative on the reimbursement plan if the tenant pays for the improvement is to have the property appraised when the tenant leaves. The landlord would pay that amount or a percentage of that amount. Of course, this agreement should also be made in advance.

Keep in mind some increased costs because of the improvement. Most building improvements will increase the assessed valuation of the property. That means more property tax. There may also be repairs to make on the improvements. And there may be extra insurance costs. For livestock buildings or grain drying and storage facilities, there may be increased costs for electricity and water.

All these increased costs should be shared in relation to how the increased income from the improvements is shared.

Let's consider that hog building described before. The tenant would pay all those extra costs if the hog income is all his. If it's a livestock share arrangement, those costs might be split according to the ratio of income and expenses agreed on in the lease. The landlord would usually pay all the costs in that case.

Another way to add facilities to a rented farm may be for the landlord to provide the materials for new buildings and the tenant either to provide his own labor for construction or hire the labor done.

Summary

You can see that there needs to be a lot of give and take in agreements involving building up a rented farm. These guidelines should help you come up with an agreement that will work for both the landlord and tenant.

When you make that agreement, just make sure it is in writing. And be sure the agreement is drawn before the investment is made. It's too easy to forget some details of an agreement you made months—or years—ago.

Chapter 15

How To Hang Onto Your Rented Land

In other businesses, they call it public relations. But in farming, you're more likely to call it "keeping the landlord happy."

However you tag it, a good relationship with your landlord can make your life easier and your farming more profitable in the long run.

Communicate

The biggest pitfall in tenant-landlord relations is often caused by poor communications. For example, one partner might get irritated because the other one didn't inform or consult with him before doing something.

The best advice on how to keep your landlord happy is to keep him well informed about what you are doing. That doesn't mean you have to get his permission before you do anything. But on the big decisions—or anything not spelled out in the lease—keep him informed.

Develop a crop plan and show it to the landlord before the crop season starts. Include a field map showing crops to be planted and chemical and fertilizer plans.

Rented land probably isn't the place to experiment with new, untried techniques. Try it first on your own land. Then if it works, share the benefits with your landlord the next year.

Be sure to keep your landlord informed about problems.

For instance, suppose you were an absentee landowner and heard that there was a planting delay in the area where you own a farm. Wouldn't you be a little nervous about whether or not your crops got planted on time? That's where a tenant might spend a few dollars for a phone call to avoid a strained relationship with the landlord.

If there's an insect problem, bad weather, or breakdowns, let the farm owner know. He'll be a lot less disappointed at harvest if he knows what happened beforehand. Otherwise, your problems may be misread as excuses. Be sure the problems aren't the result of poor management or planning on your part.

Some farm managers send their landowners a monthly report that shows costs to date and gives a summary of what has been done and how the crop looks.

The smart tenant spends a few dollars and some time for phone calls or letters to explain problems and the progress he is making. The owner will put less blame on you if you let him know other farmers are in the same boat. Keeping the landlord informed lets him know you're informed. Call when there's good news, too.

If you don't keep him up to date about problems that hurt yields, he may be shocked to find his income to be less than he expected. That can raise suspicions and cause him hard feelings. The same goes for unexpected expenses.

Teach your landlord

If your landlord is a widow or a non-farm investor, he or she may not know much about farming. Therefore, you should assume the role of an educator to help him/her understand the problems that you, as a farmer, encounter.

Since such owners often don't understand some of the common terms such as NH_3, P, K, and many of the chemical names, it might be a good idea to invite your landlord to some crop and chemical meetings conducted by the extension service or held by dealers. Or you may have to bring your landlord up to date on some modern farming practices.

That does two things. It shows the owner that you are trying to keep up to date. It also keeps him aware of the costs and other obstacles between you and good profits. It takes money to make money in farming. Be sure the landlord understands that.

Explain farm costs

Another thing that bothers many landlords is not knowing what to expect and when. For example, someone who doesn't understand farming may not understand the bills they will get before, during, and just after spring planting season. With this in mind, a sharp tenant will prepare a budget before planting to show the landlord an estimate of what the crop is going to cost. Costs or figures in this book and those put out by your extension service are averages. If you use more fertilizer to get higher yields, point it out to the landlord. He may be comparing your figures with someone else's who doesn't use enough fertilizer.

A $500 bill for ''Treflan appl.'' caught one widow off guard. She didn't know what the chemical was or why it was being used. A note to explain it would have taken only a minute and would have avoided a potential problem.

Many operators who rent from several landlords do volume buying to qualify for discounts. Then they prorate the cost back to the owners, giving them a share of the savings. But owners may wonder if they are really getting anything out of this volume discount.

You can avoid these uneasy questions by giving the owner a copy of the original invoice rather than just sending him your bill. It's easy to get photocopies of these. Your bank may even do it for you at a low cost.

Most chemical, fertilizer, and feed companies have booklets telling their rates and approximate cost per acre. Take some of these to your absentee landlord to illustrate your point.

Another good idea is for you to keep your landlord well informed on various government programs. For example, unless

someone has been following the farm papers, he may have difficulty understanding a set aside or diversion program. Make sure
your landlord gets copies of the articles. Don't hesitate to deliver
these and explain how they apply to your farm operation. You
may even want to subscribe to several farm magazines for him
so that he keeps well informed with what's going on in farming
currently.

The decision of whether or not to participate in a government
program may need to be made jointly. Pencil out the alternatives
with the landlord before making a decision.

Crop reports

The question landlords ask a tenant most often is probably,
"How are the crops doing?" As all farmers know, answering
this question accurately is very tough. Even when a crop is good
today, tomorrow wind, hailstorm, or drouth may ruin a crop. So,
you may want to make sure your landlord is immediately well informed when such things happen. Encourage the landlord to
visit the farm, especially during the growing season.

You may even want to estimate your yields a little low before
harvest so if something does happen, it doesn't seem quite so
catastrophic. Plus, if things do pan out like you think they will,
you can give your landowner a pleasant surprise in the form of a
larger check.

Keep the farm in shape

Another common hassle between the landowner and tenant is
the appearance of the property. If the landowner drives by and
spots poor weed control in the corn and soybeans or weeds in
the fence rows, the tenant will likely lose some points.

Some landlords use the farm's appearance as a measuring
stick in judging the tenant's ability. Remember that many landlords have lived on these farms themselves. So there's a certain

amount of pride in its appearance.

If you encounter a problem in cutting the weeds or painting buildings, it is wise to contact the landlord and explain why. Otherwise, if he drives by unexpectedly, you may lose his favor.

When you do have the farm looking real sharp, with the weeds all mowed and the fences and buildings in good repair, take time to invite the landlord out to look around. This may pay off in later lease negotiations.

Spending a few days during the slack seasons doing odd jobs and minor repair work on buildings will show your interest in keeping the farm in top condition. It may help you get more or better land to rent.

Chapter 16

The Farm Manager

Some farm owners don't want to—or can't—handle the management of a farm they own. So they turn those chores over to someone else.

Maybe you're a widow who doesn't have the expertise or you simply don't want to get involved in the rental or operation of the farm. Or possibly, you plan to retire somewhere else or want to pursue different interests. Yet you want to be sure the farm is well managed and that a good tenant will be selected to farm your property.

Farm management firms are in the business for just that purpose. A professional farm manager will take over your chores as the farm owner. He will be involved in detailed planning of the farm lease and will oversee every aspect of your farming operation. He will match the tenant to the farm.

The demand for this type of farm management service has grown in recent years. It's estimated that as much as 10% of the farmland in some areas of the United States is now under the care of professional farm managers.

The farm manager is a professional. He's usually an agricultural college graduate. He probably manages a number of farms, possibly as many as 40 or 50, for owners. The professional farm manager more than likely attends farm management meetings and seminars throughout the year to keep updated on current lease arrangements and farming trends. He should have a thorough knowledge of and background in cropping practices, modern technology, conservation, and marketing.

For example, a farm manager may become completely involved in the cropping plan for the farm he manages. He'll recommend and plan field arrangements and methods of application. He will also suggest tillage practices and whether certain fields should be plowed or no-till farmed to save soil. He will either make a recommendation to you or he will make the final decision for you if you prefer.

The farm manager works with the tenant to develop a plan for the farm. This is usually done well in advance of planting time. Then the manager sees that the work is done the right way and on time.

One of the most important parts of their work is marketing. A good farm manager will set up market goals with the owner and then develop a long range plan to meet those goals. That may involve forward pricing grain through a local elevator or by hedging on the futures market. But if you prefer to make the final selling decisions, you can.

Farm managers are equally involved in details of managing livestock leases. For instance, with the tenant, the manager will decide the type and number of feeder cattle to buy, their ration, health program, and marketing program.

Capital improvements

Another important role of the professional farm manager is to make capital improvement recommendations to the owner. Such improvement might include tiling, building grain bins, or remodeling the house. A program might be set up for improvements to be made over a 5 year period. Then the manager gets cost estimates and supervises the capital improvement from start to finish.

The owner, however, has the final say on what improvements are to be made.

Detailed accounting

Farm managers also keep the owner informed. They submit reports and do complete and detailed accounting for the farm. They provide the owner with monthly, semi-annual, or annual progress reports, as desired. A year end financial summary is provided for income tax purposes.

Writes the lease

The professional farm manager also develops the lease for renting the farm. His goal should be to make the lease fair to both the owner and tenant. Sometimes a manager taking over a new farm will actually write a lease more favorable to the tenant than leases in the past.

The manager wants the lease to be not only fair to both parties, but also in the best interest of the farm and owner. In some cases, that may mean a little less cash rent for the landowner, but a better tenant who will take better care of the farm. The long term payoff may still be better for the owner.

In some cases, the landlord may not have been receiving a fair rent because the lease was out of date. A new lease may result in a higher return for the landlord.

Management costs

Farm managers normally charge a percentage of the gross income for their services. The percentage may vary. But 7% to 10% of the landlord's gross income for crop share and livestock share operations is the going rate in many areas. For cash leasing, the fee also runs up to 10% of the cash rent.

Since his fee depends on production, he's going to shoot for high yields and the best market price he can get. That helps you, too. One farm management firm, for instance, claims it actually increases income enough to offset its management fee just on seed variety selection.

Selecting a farm manager

Other farm owners, your banker, farm supplier, or county agent can usually recommend professional farm managers to you. Visit with several different firms to determine if their range of services, fees, and your goals are the same.

Check out various farm managers with other landowners who use their services.

The American Society of Farm Managers and Rural Appraisers can furnish a listing of professional farm managers in your state who are members of the society. Their address is: ASFMRA, P.O. Box 6857, Denver, Colorado 80206.

Chapter 17

Material Participation

One landowner doesn't want the rent he receives to count as self employed earnings. It would cut his social security benefit.

A younger landowner wants his rent to count as self employed earnings. That way, he pays more into the social security fund and will receive bigger benefits when he retires.

Still another retired farmer isn't sure which way to go. He doesn't want his social security benefit to be cut. But he wants to have his estate qualify for current use valuation.

Material participation is the key. If you don't materially participate in the farming operation, the rent you receive is considered to be investment income. You don't pay social security tax on it. And the income doesn't cut your social security benefits if you are retired.

If you do materially participate, the rent is considered to be self employed earnings. Then you do pay social security tax on it. Plus, it can cut your social security benefits if you are retired, have earned income above the amount allowed, and are under age 72.

Whether you materially participate or not depends on the kind of rental arrangement you have. You can probably build your lease so you do materially participate. Or you can build it so you don't.

It's wise to spell out all your intentions in the written lease. Specify what the landlord is going to do to materially participate. The following guidelines will help you make sure your contributions, as landlord, are enough to qualify as materially participating if that is your goal. If you don't want to be considered to be materially participating, make sure your contributions aren't too much.

Your participation may be in either the production itself (labor or costs) or management of the production of the crops or livestock.

There are four tests to see if you are materially participating.

Test one

The landlord is considered to be materially participating if he does any three of the four following things:

- Advances, pays for, or stands good for a significant part of the cost of production.
- Furnishes a significant part of the tools, equipment, and livestock used in production.
- Makes periodic inspection of the production activities.
- Advises and consults with the tenant periodically.

There's no precise amount that can be called a significant part of the cost of production or of the tools, equipment, or livestock. One-half or more of the tools, equipment, and livestock or one-half or more of the cost of production will usually be considered significant. But an amount less than that could be considered significant depending on your own particular situation.

The costs of production are expenses that relate directly to production of commodities in the farming operation, such as feed, seed, plants, fertilizer, fuel, machinery repair, pesticides, and other supplies.

Living expenses the landlord pays for the tenant are not considered costs of production. Neither are overhead expenses such as taxes, depreciation, and insurance, for example. Also, the value of the labor furnished by the tenant cannot be counted in figuring the costs of production for this purpose.

Periodic inspection of the production activities will count toward material participation if it is done to promote production. Therefore, for a landlord's inspection activities to be counted, they must be for the purpose of seeing whether the farm work is being done properly, whether anything else needs to be done, or when it should be done.

Mere inspection to determine the condition of the buildings, fences, or other improvements does not count.

Making inspections during the soil preparation, planting, cultivating, and harvesting seasons counts even if there is no inspection during most of the growing season.

Advice and consultation will be counted toward material participation if it is done in connection with what, where, when, or how things are to be done in producing the commodities. It may be connected directly with actual planting, cultivation, and harvesting. Or, it may be connected to deciding what crops are to be planted, the kind of seed to be used, the fertility program, or the chemical control program.

Test two

A landlord may be materially participating if he regularly and frequently makes decisions which significantly affect the success of the enterprise. Making final decisions such as what, when, and where to plant, cultivate, or spray, or when to harvest the crop, what goods to buy, sell, or rent, and what records to keep will count toward material participation.

Test three

A landlord is considered to be materially participating if he works at least 100 hours over 5 or more weeks (not necessarily consecutive) on activities connected with the production of the crop.

If the landlord does less than 100 hours of work or works in less than 5 different weeks, he or she may still be materially participating if the work done adds up to a significant contribution to the production of the crop.

To be materially participating under this test, the landlord doesn't have to help with the actual plowing, planting, cultivating, and harvest, for example. The work may be making purchases, keeping the records, caring for the livestock, or repairing buildings, fences, and farm equipment used in connection with production of the crops or livestock.

However, if the landlord works as the employee of the tenant and is paid separately, that work does not count toward material participation.

Test four

The landlord's total activities may be considered as material participation even if he doesn't meet the tests described earlier. There may be different types of situations or combinations of factors that, based on the total picture, indicate material participation.

Check with social security

You can see that the material participation rules can be hard to define in a particular situation. Therefore, if you are in doubt about whether or not the landlord would be participating in a particular lease arrangement, it's wise to check it out with your closest social security office.

If you want to materially participate, it may be helpful to keep a daily diary of your activities. Record the dates you visited the farm, the hours you worked, and the type of work you did. Also, be sure to list planning sessions with the tenant.

Father-son arrangements

Probably the biggest problem with material participation arises when a farmer retires and starts drawing social security and his son continues in the farming operation.

It's sometimes hard to convince the Social Security Administration that the father isn't actively involved in the operation to the point where he is materially participating. Then, they may argue that his social security benefits should be reduced because of his earnings from the operation. That's especially true if the father continues to live on the farm and has daily contact with the farming son—or the tenant.

Currently, a person age 65 through age 71 can have $5000 of earned income from self employment or wages without losing any social security benefits. For each $2 earned above that amount, his social security benefits are reduced by $1. So it becomes quite important for the father to not be considered to be materially participating in these cases. After age 72, there's no limit on the amount you can earn and still receive your full social security benefit.

A wage arrangement where the father is employed by his son is usually the safer way to let the father continue in the operation. Even then, it's usually a good idea to explain your arrangement to the Social Security Administration and have them determine whether that arrangement would constitute material participation.

Estate planning

The 1976 Tax Reform Act included a special use valuation for farm estates that allows farmland to be valued at less than its fair market value in an estate if certain conditions are met. One of these conditions is that the owner has materially participated in the farming operation for at least 5 of the last 8 years before death.

However, it appears that a farming son, for example, or other qualified heir can meet the material participation rule for a parent.

If current use valuation is part of your estate plan, and you expect your heirs to use it, you'll want to make certain that the material participation requirements are met. The biggest problem may be if there isn't a qualified heir handling the farm business—a relative. Since the rules are quite complicated and detailed, be sure to check this out with your tax advisor.

Finally, as you set up a farm lease arrangement, study to determine whether or not you want to be considered to be materially participating in the operation. Then plan your lease arrangement accordingly.

Chapter 18

Lease Forms

The following pages contain farm lease forms you may want to use to develop a fair lease agreement. They are provided by the North Central Cooperative Extension Service.

You may want to use these forms as worksheets to determine how your final lease should be arranged.

You can obtain additional copies of these forms from your own cooperative extension service. Your state university extension service may also have lease forms available. Other sources where you may obtain forms include your own attorney, banker, or a farm manager in your area.

Finally, we encourage you to use lease forms as worksheets and as a written record of your farm lease agreement.

Cash Farm Lease

(with Flexible Provisions)

North Central Regional

Publication No. 76[1]

This **CASH FARM LEASE** form can provide the landlord and tenant with a guide for devloping an agreement to fit their individual situation. This form is not intended to take the place of legal advice pertaining to contractual relationships between the two parties. Because of the possibility that a farm operating agreement may be legally considered a partnership under certain conditions, seeking proper legal advice is recommended when developing such an agreement.

This lease is entered into this ______________ day of ______________________________________, 19______, between

______________________________________, landlord, of ______________________________________

____
(address)

______________________________________, spouse, of ______________________________________

hereafter known as "the landlord," and
____
(address)

______________________________________, tenant, of ______________________________________

____
(address)

______________________________________, spouse, of ______________________________________

hereafter known as "the tenant."
____
(address)

I. PROPERTY DESCRIPTION

The landlord hereby leases to the tenant, to occupy and use for agricultural and related purposes, the following described property:

____
____
____
____

consisting of approximately ______________ acres situated in ____ County (Counties), _________________________ (State) with all improvements thereon except as follows:

____
____
____
____

II. GENERAL TERMS OF LEASE

A. Time period covered. The provisions of this agreement shall be in effect for ________ year(s), commencing on the ________ day of __________________ ______________________, 19______. This lease shall continue in effect from year to year thereafter unless written notice of termination is given by either party to the other at least ________ days prior to expiration of this lease or the end of any year of continuation.

B. Review of lease. A written request is required for a general review of the lease or for consideration of proposed changes by either party, at least ________ days prior to the final date for giving notice to terminate the lease as specified in IIA.

C. Amendments and alterations. Amendments and alterations to this lease shall be in writing and shall be signed by both the landlord and tenant.

D. No partnership intended. It is particularly understood and agreed that this lease shall not be deemed to be nor intended to give rise to a partnership relation.

E. Transfer of property. If the landlord should sell or otherwise transfer title to the farm, he will do so subject to the provisions of this lease.

F. Right of entry. The landlord reserves the right for himself, his agents, his employees, or his assigns to enter the farm at any reasonable time to: a) consult with the tenant; b) make repairs, improvements, and inspections; and c) (after notice of termination of the lease is given) do plowing, seeding, fertilizing, and any other customary seasonal work, none of which is to interfere with the tenant in carrying out regular farm operations.

G. No right to sublease. The landlord does not convey to the tenant the right to lease or sublease any part of the farm or to assign the lease to any person or persons whomsoever.

H. Binding on heirs. The provisions of this lease shall be binding upon the heirs, executors, administrators, and successors of both landlord and tenant in like manner as upon the original parties, except as provided by mutual written agreement.

I. Additional provisions.

____
____
____
____
____

III. LAND USE

A. General provisions. The land described in Section I will be used in approximately the following manner. If it is impracticable in any year to follow such a land use plan, appropriate adjustments will be made by mutual agreement between the parties.

1. Cropland

a) Row crops	____________ Acres
b) Small grains	____________ Acres
c) Legumes	____________ Acres
d) Rotation pasture	____________ Acres
2. Permanent pasture	____________ Acres

3. Other:

_______________________	_____________ Acres
_______________________	_____________ Acres
_______________________	_____________ Acres
4. Total	_____________ Acres

B. Restrictions. The maximum acres harvested as silage shall be ________ acres unless it is mutually decided otherwise.
The pasture stocking rate shall not exceed:

PASTURE IDENTIF.

_______________________	_____________ acres/animal unit
_______________________	_____________ acres/animal unit
_______________________	_____________ acres/animal unit

(1000 pound mature cow is equivalent to one animal unit.)

Other restrictions are:

____
____
____
____

C. Government programs. The extent of participation in government programs will be discussed and decided on an annual basis. The course of action agreed upon shall be placed in writing and be signed by both parties. A copy of the course of action so agreed upon shall be made available to each party.

IV. AMOUNT AND PAYMENT OF RENT

(If a flexible cash rental arrangement is desired, use material on the last page of this form and omit section A below.)

A. Cash rental rates. The tenant agrees to pay as cash rent the amount as calculated below for each kind of land; or, one total may be entered for ENTIRE FARM UNIT.

Amount of Cash Rent

Kind of Land or Improvements	Acres	Rate/ Acre	Amount
Row Crops		$	$
Small Grains		$	$
Legumes		$	$
Permanent Pasture		$	$
Timber		$	$
Waste		$	$
Farm Buildings	XXXXX	XXXXX	$
Dwelling	XXXXX	XXXXX	$
Other			$
ENTIRE FARM UNIT		$XXXX	$

B. Rental payment. The annual cash rent shall be paid as follows:

$________ on or before ______ day of ________ (month),
$________ on or before ______ day of ________ (month),
$________ on or before ______ day of ________ (month),
$________ on or before ______ day of ________ (month),

If rent is not paid when due, the tenant agrees to pay interest on the amount of unpaid rent at the rate of ________ percent per annum from the due date until paid.

C. Rental adjustment—Additional agreements in regard to rental payment:

____
____

V. OPERATION AND MAINTENANCE OF FARM

In order to operate this farm efficiently and to maintain it in a high state of productivity, the parties agree as follows:

A. The tenant agrees:

1. General maintenance. To provide the unskilled labor necessary to maintain the farm and its improvements during his tenancy in as good condition as it was at the beginning. Normal wear and depreciation and damage from causes beyond the tenant's control are excepted.

2. Land use. Not to: a) plow permanent pasture or meadowland, b) cut live trees for sale or personal uses, or c) pasture new seedings of legumes and grasses in the year they are seeded without consent of the landlord.

3. Insurance. Not to house automobiles, motortrucks, or tractors in barns, or otherwise violate restrictions in the landlord's insurance policies without written consent from the landlord. Restrictions to be observed are as follows:

____
____
____
____

4. Noxious weeds. To use diligence to prevent noxious weeds from going to seed on the farm. Treatment of the noxious weed infestation and cost thereof shall be handled as follows:

____
____
____

5. Addition of improvements. Not to: a) erect or permit to be erected on the farm any nonremovable structure or building, b) incur any expense to the landlord for such purposes, or c) add electrical wiring, plumbing, or heating to any building without written consent of the landlord.

6. Conservation. Control soil erosion as completely as practicable; keep in good repair all terraces, open ditches, inlets and outlets of tile drains; preserve all established watercourses or ditches including grassed waterways; and refrain from any operation or practice that will injure such structures.

7. Damages. When he leaves the farm, to pay the landlord reasonable compensation for any damages to the farm for which he, the tenant, is responsible. Any decrease in value due to ordinary wear and depreciation or damages outside the control of the tenant are excepted.

8. Costs of operation. To pay all costs of operation except those specifically referred to in Sections V-A-4 and V-B.

9. Repairs. Not to buy materials for maintenance and repairs in an amount in excess of $____________ within a single year without written consent of the landlord.

B. The landlord agrees:

1. Loss replacement. To replace or repair as promptly as possible the dwelling or any other building regularly used by the tenant that may be destroyed or damaged by fire, flood, or other cause beyond the control of the tenant or to make rental adjustments in lieu of replacements.

2. Materials for repairs. To furnish all material needed for normal maintenance and repairs.

3. Skilled labor. To furnish any skilled labor for tasks which the tenant himself is unable to make satisfactorily. Additional agreements regarding materials and labor are:

____

____

____

4. Reimbursement. To pay for materials purchased by the tenant for purposes of repair and maintenance in an amount not to exceed $__________ in any one year, except as otherwise agreed upon. Reimbursement shall be made within ________ days after the tenant submits the bill.

5. Removable improvements. Let the tenant make minor improvements of a temporary or removable nature, which do not mar the condition or appearance of the farm, at the tenant's expense. He further agrees to let the tenant remove such improvements even though they are legally fixtures at any time this lease is in effect or within __________ days thereafter, provided the tenant leaves in good condition that part of the farm from which such improvements are removed. The tenant shall have no right to compensation for improvements that are not removed except as mutually agreed.

6. Compensation for crop expenses. To reimburse the tenant at the termination of this lease for field work done and for other crop costs incurred for crops to be harvested during the following year. Unless otherwise agreed, current custom rates for operations involved will be used as a basis of settlement.

C. Both agree:

1. Not to obligate other party. Neither party hereto shall pledge the credit of the other party hereto for any purpose whatsoever without the consent of the other party. Neither party shall be responsible for debts or liabilities incurred, or for damages caused by the other party.

2. Capital improvements. That costs of establishing hay or pasture seedings, new conservation structures, improvements (except as provided in Section V-B-5), or of applying lime and other long-lived fertilizers shall be divided between landlord and tenant as set forth in the following table. The tenant will be reimbursed by the landlord either when the improvement is completed, **or** the tenant will be compensated for his share of the depreciated cost of his contribution when he leaves the farm based on the value of the tenant's contribution and depreciation rate shown in Table I. (Cross out the portion of the preceding sentence which does not apply).

   Rates for labor, power, and machinery contributed by the tenant shall be agreed upon before construction is started.

Table I—Compensation for Improvements

| Type of Improvement | Date to Be Completed | Estimated Total Cost (dollars) | Proportion to be Contributed by tenant | | | Total Value of Tenant's Contrib. (dollars) * | Rate of Annual Depreciation |
			Mat.	Unskilled Labor	Mach.		
			%	%	%		%
		$				$	
		$				$	
		$				$	
		$				$	
		$				$	
		$				$	
		$				$	
		$				$	
		$				$	
		$				$	
		$				$	
		$				$	
		$				$	
		$				$	
		$				$	
		$				$	

* To be recorded when improvement is completed.

VI. ARBITRATION OF DIFFERENCES

Any differences between the parties as to their several rights or obligations under this lease that are not settled by mutual agreement after thorough discussion, shall be submitted for arbitration to a committee of three disinterested persons, one selected by each party hereto and the third by the two thus selected. The committee's decision shall be accepted by both parties.

AMOUNT OF RENT TO BE PAID WHEN CROPLAND IS RENTED ON A FLEXIBLE BASIS.

A. Cash Rent for Non-Flexible Items (complete at beginning of lease period).

    a. Pasture .... $___________

    b. Hayland .... $___________

    c. Other Non-Flexible Cropland .... $___________

    d. Timber-Wasteland .... $___________

    e. Farmstead .... $___________

         TOTAL NON-FLEXIBLE RENT .... $___________

B. Flexible Cropland Rent (From Method I, II, or III below .... $___________

C. TOTAL RENT FOR YEAR .... $___________

D. Flexible Cropland Rent (use Method I, II, or III).

    1. BASIC INFORMATION TO BE USED IN METHODS I AND II.

Crop(s)	Base Cash Rent (per acre)	Base Yield (bu. or ton per acre)	Base Price (per bu. or per ton)	Minimum Cash Rent (per acre)	Maximum Cash Rent (per acre)
_________	$_________	_________	$_________	$_________	$_________
_________	$_________	_________	$_________	$_________	$_________
_________	$_________	_________	$_________	$_________	$_________

    2. THE CURRENT PRICE FOR THE CURRENT YEAR SHALL BE AVERAGE PRICE AT CLOSE OF DAY BASED ON THE FOLLOWING TIME PERIOD(S) AND LOCATION(S):

Crop(s)   Price Source

_________ _____ Day _____ Month through _____ Day _____ Month at _______________

_________ _____ Day _____ Month through _____ Day _____ Month at _______________

_________ _____ Day _____ Month through _____ Day _____ Month at _______________

FOR EACH YEAR OF THIS LEASE, THE PER ACRE BASE CASH RENT FOR EACH CROP SPECIFIED SHALL BE ADJUSTED AT THE CLOSE OF THE CROPPING SEASON BY ONE OF THE FOLLOWING METHODS:

METHOD I—FLEXING FOR PRICE ONLY.

$$\text{Crop(s)} : \text{Base Rent} \times \frac{\text{Current Pr.}}{\text{Base Pr.}} = \text{Rent/Ac.}^1 \times \text{Acres Grown} = \text{Adj. Rent for Yr.}$$

_________ $_________ X ($/$ _________) = $_________ X _________ = $_________

_________ $_________ X ($/$ _________) = $_________ X _________ = $_________

_________ $_________ X ($/$ _________) = $_________ X _________ = $_________

                                            Total all crops = $_________

METHOD II—FLEXING FOR PRICE AND YIELD.

$$\text{Crop(s)} : \text{Base Rent} \times \frac{\text{Current Pr.}}{\text{Base Pr.}} \times \frac{\text{Cur. Yld.}^2}{\text{Base Yld.}} = \text{Rent/Ac.}^1 \times \text{Acres Grown} = \text{Adj. Rent for Yr.}$$

_________ : $_________ X ($/$ _________) X _________ = $_________ X _________ = $_________

_________ : $_________ X ($/$ _________) X _________ = $_________ X _________ = $_________

_________ : $_________ X ($/$ _________) X _________ = $_________ X _________ = $_________

                                            Total all crops = $_________

1. If calculated figure is less than Minimum Cash Rent in D-1, use the Minimum. If calculated figure is more than Maximum Cash Rent in D-1, use the Mamixum.
2. The current yield shall be the "farm" yield for the current lease year.

METHOD III—WORK OUT AND RECORD PROCEDURE TO BE USED.

____

____

____

____

Executed in duplicate on the date first above written:

_______________________________________   _______________________________________
(tenant)   (landlord)

_______________________________________   _______________________________________
(tenant spouse)   (landlord spouse)

COUNTY OF ___________________________________

STATE OF ___________________________________   } SS:

On this _________________________ day of _________________________ A.D., 19_____, before me, the undersigned, a Notary Public in said State, personally appeared _________________________________

_________________________________, _________________________________, _________________________________,

and _________________________________, to me known to be the identical persons named in and who executed the foregoing instrument, and acknowledged that they executed the same as their voluntary act and deed.

____
Notary Public

Issued in furtherance of Cooperative Extension work, Acts of Congress of May 8 and June 30, 1914, in cooperation with the U.S. Department of Agriculture and Cooperative Extension Services of Illinois, Indiana, Kansas, Michigan, Minnesota, Missouri, Nebraska, North Dakota, Ohio, South Dakota and Wisconsin. John O. Dunbar, Director, Cooperative Extension Service, Kansas State University, Manhattan.

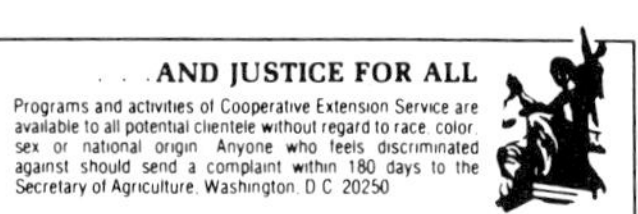

Crop-Share or Crop-Share-Cash Farm Lease

North Central Regional
Publication No. 77

This form can provide the landlord and tenant with a guide for developing an agreement to fit their individual situation. This form is not intended to take the place of legal advice pertaining to contractual relationships between the two parties. Because of the possibility that a farm operating agreement may be legally considered a partnership under certain conditions, seeking proper legal advice is recommended when developing such an agreement.

This lease is entered into this _____________ day of ____, 19______, between

____, landlord, of ____

____
(address)

____, spouse, of ____

____
(address)

hereafter known as "the landlord," and

____, tenant, of ____

____
(address)

____, spouse, of ____

____
(address)

hereafter known as "the tenant."

I. PROPERTY DESCRIPTION

The landlord hereby leases to the tenant, to occupy and use for agricultural and related purposes, the following described property:

____

____

____

consisting of approximately __________ acres situated in ____ County (Counties), ____ (State) with all improvements thereon except as follows:

____

____

____

II. GENERAL TERMS OF LEASE

A. Time period covered. The provisions of this agreement shall be in effect for ________ year(s), commencing on the ______ day of __________________, 19______. This lease shall continue in effect from year to year thereafter unless written notice of termination is given by either party to the other at least ________ days prior to expiration of this lease or the end of any year of continuation.

B. Review of lease. A written request is required for a general review of the lease or for consideration of proposed changes by either party, at least ________ days prior to the final date for giving notice to terminate the lease as specified in IIA.

C. Amendments and alterations. Amendments and alterations to this lease shall be in writing and shall be signed by both the landlord and tenant.

D. No partnership intended. It is particularly understood and agreed that this lease shall not be deemed to be nor intended to give rise to a partnership relation.

E. Transfer of property. If the landlord should sell or otherwise transfer title to the farm, he will do so subject to the provisions of this lease.

F. Right of entry. The landlord reserves the right for himself, his agents, his employees, or his assigns to enter the farm at any reasonable time to: a) consult with the tenant; b) make repairs, improvements, and inspections; and c) (after notice of termination of the lease is given) do plowing, seeding, fertilizing, and any other customary seasonal work, none of which is to interfere with the tenant in carrying out regular farm operations.

G. No right to sublease. The landlord does not convey to the tenant the right to lease or sublet any part of the farm or to assign the lease to any person or persons whomsoever.

H. Binding on heirs. The provisions of this lease shall be binding upon the heirs, executors, administrators, and successors of both landlord and tenant in like manner as upon the original parties, except as provided by mutual written agreement.

I. Landlord's lien for rent and performance. The landlord's lien provided by law on crops grown or growing shall be the security for the rent herein specified and for the faithful performance of the terms of the lease. If the tenant fails to pay the rent due or fails to keep the agreements of this lease, all costs and attorney fees of the landlord in enforcing collection or performance shall be added to and become a part of the obligations payable by the tenant hereunder.

J. Additional provisions.

____

____

III. LAND USE

A. General provisions. The land described in Section I will be used in approximately the following manner. If it is impractical in any year to follow such a land-use plan, appropriate adjustments will be made by mutual written agreement between the parties.

1. Cropland
   a) Row crops   __________ acres
   b) Small grains   __________ acres
   c) Legumes   __________ acres
   d) Rotation pasture   __________ acres
2. Permanent pasture: __________ acres
3. Other: __________________ __________ acres
   __________________________ __________ acres
4. Total   __________ acres

B. Restrictions. The maximum acres harvested as silage shall be _______ acres unless it is mutually decided otherwise.
The pasture stocking rate shall not exceed:

PASTURE IDENTIF.

__________________	__________ acres/animal unit
__________________	__________ acres/animal unit
__________________	__________ acres/animal unit

(1000 pound mature cow is equivalent to one animal unit.)

Other restrictions are:

____

____

____

____

____

C. Government programs. The extent of participation in government programs will be discussed and decided on an annual basis. The course of action agreed upon shall be placed in writing and be signed by both parties. A copy of the course of action so agreed upon shall be made available to each party.

IV. CROP-SHARE-CASH RENT AND RELATED PROVISIONS

A. General agreement. The tenant agrees to pay as rent for the use of the land the share of crops shown in Table 1 of this section. The tenant also agrees to furnish all labor, machinery, and cash operating expenses except for landlord's share (percent and/or dollar charge per unit) indicated in Table 1.

Table 1—Landlord's Share (% and/or $) of Crops and Crop Expenses

	Corn example	Corn	Grain sorghum	Small grain	Soybeans	Hay, ______		
SHARE OF CROPS	50%							
SHARE OF CROP EXPENSES:								
Fertilizer:								
Materials	50%							
Application	50%							
Herbicide:								
Materials	50%							
Application								
Insecticide:								
Materials	50%							
Application								
Seed	50%							
Lime, rock phosphate*	100%							
Harvesting (per ac.)	$7.50							
Drying	50%							
Baling								
Delivery to:								
Storage/bu.								
Market/bu.	$ .07							

* Lime, rock phosphate, and other fertilizers having more than one year life paid by the tenant should be recorded in the compensation table in Section V-C-2.

B. Other crop-share-cash agreements.

1. Operating expenses. Additional agreements relative to the sharing of expenses are as follows:

____

____

____

2. Storage, landlord's crop. At the landlord's request, the tenant agrees to store as much of the landlord's share of the crops as possible, using storage space reserved by the landlord and not to exceed _______ . percent of the storage space not specifically reserved.

3. Delivery of grain. The tenant agrees to deliver the landlord's share of crops at a place and at a time the landlord shall designate, not over __________ miles distant at the charge shown in Table 1 of this section. Additional agreements are:

____

____

____

4. Cash rent on non-shared items. The tenant agrees to pay cash rent annually for the use of the following non-shared items.

Table 2—Amount of Annual Cash Rent
(Complete at beginning of lease)

	Total
Pasture	$__________
Hayland: ______________________	$__________
______________________	$__________
Farmstead: Dwelling	$__________
Service bldgs.	$__________
Timber and waste	$__________
Total cash rent	$__________

Payment of cash rent: The tenant agrees to pay cash rent as follows:

$________ on or before ______ day of ______ (month)

$________ on or before ______ day of ______ (month)

$________ on or before ______ day of ______ (month)

$________ on or before ______ day of ______ (month)

If rent is not paid when due, the tenant agrees to pay interest on the amount of unpaid rent at the rate of ________ percent per annum from due date until paid.

5. Pasturing. The tenant will prevent damage to cropland and growing crops by livestock.

6. Home use. The tenant and landlord may take for home use the following kinds and quantities of jointly owned crops:

____

____

____

____

7. Buying and selling. The landlord and tenant will buy and sell jointly owned property according to the following agreement:

____

____

____

____

8. Division of property. At the termination of this lease, all jointly owned property will be divided or disposed of as follows:

____

____

____

V. OPERATION AND MAINTENANCE OF FARM

In order to operate this farm efficiently and to maintain it in a high state of productivity, the parties agree as follows:

A. The tenant agrees:

1. General maintenance. To provide the unskilled labor necessary to maintain the farm and its improvements during his tenancy in as good condition as it was at the beginning. Normal wear and depreciation and damage from causes beyond the tenant's control are excepted.

2. Land use. Not to: a) plow pasture or meadowland, b) cut live trees for sale or personal use, or c) pasture new seedings of legumes and grasses in the year they are seeded without consent of the landlord.

3. Insurance. Not to house automobiles, motor trucks, or tractors in barns, or otherwise violate restrictions in the landlord's insurance policies without written consent from the landlord. Restrictions to be observed are as follows:

____

____

____

4. Noxious weeds. To use diligence to prevent noxious weeds from going to seed on the farm. Treatment of the noxious weed infestation and cost thereof shall be handled as follows:

____

____

____

5. Addition of improvements. Not to: a) erect or permit to be erected on the farm any nonremovable structure or building, b) incur any expense to the landlord for such purposes, or c) add electrical wiring, plumbing, or heating to any building without written consent of the landlord.

6. Conservation. Control soil erosion as completely as practicable; keep in good repair all terraces, open ditches, inlets and outlets of tile drains; preserve all established watercourses or ditches including grassed waterways; and refrain from any operation or practice that will injure such structures.

7. Damages. When he leaves the farm, to pay the landlord reasonable compensation for any damages to the farm for which he, the tenant, is responsible. Any decrease in value due to ordinary wear and depreciation or damages outside the control of the tenant are excepted.

8. Costs of operation. To pay all costs of operation except those specifically referred to in Sections IV, V-A-4, and V-B.

9. Repairs. Not to buy materials for maintenance and repairs in an amount in excess of $____________ within a single year without written consent of the landlord.

B. The landlord agrees:

1. Loss replacement. To replace or repair as promptly as possible the dwelling or any other building regularly used by the tenant that may be destroyed or damaged by fire, flood, or other cause beyond the control of the tenant or to make rental adjustments in lieu of replacements.

2. Materials for repairs. To furnish all material needed for normal maintenance and repairs.

3. Skilled labor. To furnish any skilled labor tasks which the tenant himself is unable to perform satisfactorily. Additional agreements regarding materials and labor are:

____

____

____

4. Reimbursement. To pay for materials purchased by the tenant for purposes of repair and maintenance in an amount not to exceed $____________ in any one year, except as otherwise agreed upon. Reimbursement shall be made within ________ days after the tenant submits the bill.

163

5. Removable improvements. Let the tenant make minor improvements of a temporary or removable nature, which do not mar the condition or appearance of the farm, at the tenant's expense. He further agrees to let the tenant remove such improvements even though they are legally fixtures at any time this lease is in effect or within ______ days thereafter, provided the tenant leaves in good condition that part of the farm from which such improvements are removed. The tenant shall have no right to compensation for improvements that are not removed except as mutually agreed.

6. Compensation for crop expenses. To reimburse the tenant at the termination of this lease for field work done and for other crop costs incurred for crops to be harvested during the following year. Unless otherwise agreed, current custom rates for the operations involved will be used as a basis of settlement.

C. Both agree:

1. Not to obligate other party. Neither party hereto shall pledge the credit of the other party hereto for any purpose whatsoever without the consent of the other party. Neither party shall be responsible for debts or liabilities incurred, or for damages caused by the other party.

2. Capital improvements. Costs of establishing hay or pasture seedings, new conservation structures, improvements (except as provided in Section V-B-5), or of applying lime and other longlived fertilizers shall be divided between landlord and tenant as set forth in the following table. The tenant will be reimbursed by the landlord either when the improvement is completed, **or** the tenant will be compensated for his share of the depreciated cost of his contribution when he leaves the farm based on the value of the tenant's contribution and depreciation rate shown in the following table. (Cross out the portion of the preceding sentence which does not apply.)

Rates for labor, power, and machinery contributed by the tenant shall be agreed upon before construction is started.

3. Mineral rights. Nothing in this lease shall confer upon the tenant any right to minerals underlying said land, but same are hereby reserved by the landlord together with the full right to enter upon the premises and to bore, search, and excavate for same, to work and remove same, and to deposit excavated rubbish, and with full liberty to pass over said premises with vehicles and lay down and work any railroad track or tracks, tanks, pipelines, power lines, and structures as may be necessary or convenient for the above purpose. The landlord agrees to reimburse the tenant for any actual damage he may suffer for crops destroyed by these activities and to release the tenant from obligation to continue farming this property when development of mineral resources interferes materially with the tenant's opportunity to make a satisfactory return.

VI. ARBITRATION OF DIFFERENCES

Any differences between the parties as to their several rights or obligations under this lease that are not settled by mutual agreement after thorough discussion, shall be submitted for arbitration to a committee of three disinterested persons, one selected by each party hereto and the third by the two thus selected. The committee's decision shall be accepted by both parties.

Compensation for Improvements Table

Type of improvement	Date to be completed	Estimated total cost (dollars)	Proportion to be contributed by tenant Unskilled			Total value of tenant's contrib. (dollars) *	Rate of annual depreciation
			Material	labor	Mach.		
			%	%	%		%
		$				$	
		$				$	
		$				$	
		$				$	
		$				$	

* To be recorded when improvement is completed.

Executed in duplicate on the date first above written:

____
(tenant)

____
(tenant spouse)

____
(landlord)

____
(landlord spouse)

STATE OF _______________________
COUNTY OF _____________________ } SS:

On this _________________________ day of _________________________ A.D., 19_____, before me, the undersigned, a Notary Public in said State, personally appeared _________________________

_________________________, _________________________, _________________________,

and _________________________, to me known to be the identical persons named in and who executed the foregoing instrument, and acknowledged that they executed the same as their voluntary act and deed.

____
Notary Public

Issued in furtherance of Cooperative Extension work, Acts of Congress of May 8 and June 30, 1914, in cooperation with the U.S. Department of Agriculture and Cooperative Extension Services of Illinois, Indiana, Kansas, Michigan, Minnesota, Missouri, Nebraska, North Dakota, Ohio, South Dakota and Wisconsin. John O. Dunbar, Director, Cooperative Extension Service, Kansas State University, Manhattan.

Livestock Share Farm Lease
North Central Regional

Publication No. 108

This form can provide the landlord and tenant with a guide for developing an agreement to fit their individual situation. This form is not intended to take the place of legal advice pertaining to contractual relationships between the two parties. Because of the possibility that a farm operating agreement may be legally considered a partnership under certain conditions, seeking proper legal advice is recommended when developing such an agreement.

This lease is entered into this ______________ day of ____, 19______, between ____, landlord, of ____

(address)

____, spouse, of ____

(address)

hereafter known as "the landlord," and

____, tenant, of ____

(address)

____, spouse, of ____

(address)

hereafter known as "the tenant."

I. PROPERTY DESCRIPTION

The landlord hereby leases to the tenant, to occupy and use for agricultural and related purposes, the following described property:

____
____
____

consisting of approximately ______________ acres situated in ____________________________ County (Counties), ____________________________________ (State) with all improvements thereon except as follows:

____
____
____

II. GENERAL TERMS OF LEASE

A. Time period covered. The provisions of this agreement shall be in effect for ________ year(s), commencing on the ______________________________ day of ______________________________, 19______. This lease shall continue in effect from year to year thereafter unless written notice of termination is given by either party to the other at least ______________ days prior to expiration of this lease or the end of any year of continuation.

B. Review of lease. A written request is required for a general review of the lease or for consideration of proposed changes by either party, at least ______________ days prior to the final date for giving notice to terminate the lease as specified in IIA.

C. Amendments and alterations. Amendments and alterations to this lease shall be in writing and shall be signed by both the landlord and tenant.

D. No partnership intended. It is particularly understood and agreed that this lease shall not be deemed to be nor intended to give rise to a partnership relation.

E. Transfer of property. If the landlord should sell or otherwise transfer title to the farm, he will do so subject to the provisions of this lease.

F. Right of entry. The landlord reserves the right for himself, his agents, his employees, or his assigns to enter the farm at any reasonable time to: a) consult with the tenant; b) make repairs, improvements, and inspections; and c) (after notice of termination of the lease is given) do plowing, seeding, fertilizing, and any other customary seasonal work, none of which is to interfere with the tenant in carrying out regular farm operations.

G. No right to sublease. The landlord does not convey to the tenant the right to lease or sublet any part of the farm or to assign the lease to any person or persons whomsoever.

H. Binding on heirs. The provisions of this lease shall be binding upon the heirs, executors, administrators, and successors of both landlord and tenant in like manner as upon the original parties, except as provided by mutual written agreement.

I. Landlord's lien for rent and performance. The landlord's lien provided by law on crops grown or growing shall be the security for the rent herein specified and for the faithful performance of the terms of the lease. If the tenant fails to pay the rent due or fails to keep the agreements of this lease, all costs and attorney fees of the landlord in enforcing collection or performance shall be added to and become a part of the obligations payable by the tenant hereunder.

J. Additional provisions.

____
____

III. LAND USE

A. General provisions. The land described in Section I will be used in approximately the following manner. If it is impractical in any year to follow such a land-use plan, appropriate adjustments will be made by mutual written agreements between the parties.

1. Cropland

    a) Row crops   _____________ acres

    b) Small grains   _____________ acres

    c) Legumes   _____________ acres

    d) Rotation pasture _____________ acres

2. Pasture: _________________________ _____________ acres

3. Other: ___________________________ _____________ acres

    ___________________________________ _____________ acres

4. Total   _____________ acres

B. Restrictions. The maximum acres harvested as silage shall be _________ acres unless it is mutually decided otherwise.

The pasture stocking rate shall not exceed:

PASTURE IDENTIFICATION:

_____________________________ _________ acres/animal unit

_____________________________ _________ acres/animal unit

_____________________________ _________ acres/animal unit

        (1,000 pound mature cow is equivalent to
                 one animal unit.)

Other restrictions are:

____

____

____

C. Government programs. The extent of participation in government programs will be discussed and decided on an annual basis. The course of action agreed upon shall be placed in writing and be signed by both parties. A copy of the course of action so agreed upon shall be made available to each party.

IV. LIVESTOCK PRODUCTION AND SHARING ARRANGEMENTS

A. It is agreed the tenant and landlord will engage in the production of livestock. Real property including land and the type and number of livestock to be contributed to production by each party are reported in Table 1.

Table 1—Contributions of Property to be Furnished by Each Party

	Approximate number to be kept	Share furnished by Landlord %	Share furnished by Tenant %
1. Land and fixed improvements at beginning of this lease described in Section I		__________	__________
2. Fixed improvements constructed during period of this lease:			
Materials and skilled labor		__________	__________
Hauling materials to farm		__________	__________
Farm labor		__________	__________
__________________		__________	__________
__________________		__________	__________
3. Livestock:			
Breeding: __________	__________	__________	__________
__________________	__________	__________	__________
__________________	__________	__________	__________
Replacements: __________	__________	__________	__________
__________________	__________	__________	__________
__________________	__________	__________	__________
Feeders: __________	__________	__________	__________
__________________	__________	__________	__________
__________________	__________	__________	__________
4. Machinery and equipment: (crop, livestock, etc.)			
__________________		__________	__________
__________________		__________	__________
__________________		__________	__________
__________________		__________	__________
__________________		__________	__________
5. Portable farm buildings:			
__________________		__________	__________
__________________		__________	__________

B. Annual operating expenses shall be supplied by the landlord and tenant as reported in Table 2 except as discussed in Section VI.

Table 2—Percentage Share of Operating Expenses to be Furnished by Each Party

	(L)	(T)
1. Crop Expenses:		
• Fertilizer	_______	_______
• Lime	_______	_______
• Seed	_______	_______
• Herbicide	_______	_______
• Crop insurance	_______	_______
• Other supplies	_______	_______
• Other: _______	_______	_______
2. Livestock Expenses:		
• Feed purchased	_______	_______
• Veterinary	_______	_______
• Breeding fees	_______	_______
• Medicines and drugs	_______	_______
• Feed grinding and mixing	_______	_______
• _______	_______	_______
• _______	_______	_______
3. Fuel:		
• Tractor	_______	_______
• Truck	_______	_______
• Harvesting	_______	_______
• Crop drying	_______	_______
• Feed processing	_______	_______
• Heating buildings	_______	_______
4. Electricity	_______	_______
5. Telephone	_______	_______
6. General hired labor	_______	_______
7. Custom:		
• Hauling crops and livestock	_______	_______
• Harvesting:		
Corn	_______	_______
Small grain	_______	_______
Soybeans	_______	_______
_______	_______	_______
_______	_______	_______
8. Insurance:		
• Buildings	_______	_______
• _______	_______	_______
9. Taxes	_______	_______
10. Interest:		
• Operating capital	_______	_______
• Intermediate term loans	_______	_______
11. Other	_______	_______

C. Additional agreements in regard to livestock production.

  1. Breeding replacements shall be furnished as follows:

    ____

    ____

  2. Sale of breeding stock shall be shared as follows:

    ____

    ____

  3. Other breeding stock provisions are:

    ____

    ____

D. Neither landlord or tenant shall have the authority to bind the other in any contract with third parties. Expenses other than those reported in Tables 1 and 2 shall be shared as follows:

____

____

E. Buying and selling. The tenant shall consult with the landlord regarding time, price, sales agency, and similar matters regarding the purchase and sale of livestock, feed, and crops whenever the transaction exceeds $_______ in value. Additional agreements are as follows:

____

____

F. Livestock restrictions. Neither the tenant nor the landlord shall bring to the farm livestock not included in the agreement without express written permission of the other party.

  Additional agreements relative to livestock are:

____

____

G. Equipment and machinery replacements. The cost of additional and replacement livestock equipment and machinery will be shared as follows:

____

____

____

____

V. DIVISION OF INCOME AND CASH RENT ON NON-SHARED ITEMS

A. Division of income. The tenant shall pay rent to the landlord for the use of the landlord's property described in this lease (Table 1) an amount equal to _______ percent of the gross income. Gross income shall consist of the proceeds from the sale or exchange of all grain, forages, livestock, and other products produced under the provisions of this lease, except for:

____

____

____

B. Cash rent on non-shared items. The tenant agrees to pay cash rent annually for the use of the following non-shared items:

Table 3—Amount of Annual Cash Rent
(complete at beginning of lease)

Farmstead: Dwelling	$__________
Service buildings	$__________
Timber and waste	$__________
Other: __________________	$__________
__________________	$__________
Total cash rent	$__________

Payment of cash rent: The tenant agrees to pay cash rent as follows:

$________ on or before ______ day of ______ month

$________ on or before ______ day of ______ month

$________ on or before ______ day of ______ month

$________ on or before ______ day of ______ month

If rent is not paid when due, the tenant agrees to pay interest on the amount of unpaid rent at the rate of ________ percent per annum from due date until paid.

VI. OPERATION AND MAINTENANCE OF FARM

In order to operate this farm efficiently and to maintain it in a high state of productivity, the parties agree as follows:

A. The tenant agrees:

1. General maintenance. To provide the unskilled labor necessary to maintain the farm and its improvements during his tenancy in as good condition as it was at the beginning. Normal wear and depreciation and damage from causes beyond the tenant's control are excepted.

2. Land use. Not to: a) plow pasture or meadowland, b) cut live trees for sale or personal use, c) pasture new seedlings of legumes and grasses in the year they are seeded without consent of the landlord.

3. Insurance. Not to house automobiles, motor trucks, or tractors in barns, or otherwise violate restrictions in the landlord's insurance policies without written consent from the landlord. Restrictions to be observed are as follows:

____

____

____

4. Noxious weeds. To use diligence to prevent noxious weeds from going to seed on the farm. Treatment of the noxious weed infestation and cost thereof shall be handled as follows:

____

____

____

____

5. Addition of improvements. Not to: a) erect or permit to be erected on the farm any nonremovable structure or building, b) incur any expense to the landlord for such purposes, or c) add electrical wiring, plumbing, or heating to any building without written consent of the landlord.

6. Conservation. Control soil erosion as completely as practicable; keep in good repair all terraces, open ditches, inlets and outlets of tile drains; preserve all established watercourses or ditches including grassed waterways; and refrain from any operation or practice that will injure such structures.

7. Damages. When he leaves the farm, to pay the landlord reasonable compensation for any damages to the farm for which he, the tenant, is responsible. Any decrease in value due to ordinary wear and depreciation or damages outside the control of the tenant are excepted.

8. Costs of operation. To pay all costs of operation except those specifically referred to in Sections IV, VI-A-4, and VI-B.

9. Repairs. Not to buy materials for maintenance and repairs in an amount in excess of $____________ within a single year without written consent of the landlord.

B. The landlord agrees:

1. Loss replacement. To replace or repair as promptly as possible the dwelling or any other building regularly used by the tenant that may be destroyed or damaged by fire, flood, or other cause beyond the control of the tenant or to make rental adjustments in lieu of replacements.

2. Materials for repairs. To furnish all material needed for normal maintenance and repair.

3. Skilled labor. To furnish any skilled labor for tasks which the tenant himself is unable to perform satisfactorily. Additional agreements regarding materials and labor are:

____

4. Reimbursement. To pay for materials purchased by the tenant for purposes of repair and maintenance in an amount not to exceed $__________ in any one year, except as otherwise agreed upon. Reimbursement shall be made within ________ days after the tenant submits the bill.

5. Removable improvements. To let the tenant make minor improvements of a temporary or removable nature, which do not mar the condition or appearance of the farm, at the tenant's expense. He further agrees to let the tenant remove such improvements even though they are legally fixtures at any time this lease is in effect or within ________ days thereafter, provided the tenant leaves in good condition that part of the farm from which such improvements are removed. The tenant shall have no right to compensation for improvements that are not removed except as mutually agreed.

6. Compensation for crop expenses. To reimburse the tenant at the termination of this lease for field work done and for other crop costs incurred for crops to be harvested during the following year. unless otherwise agreed, current custom rates for the operations involved will be used as a basis of settlement.

168

C. Both agree:

1. Capital improvements. Costs of establishing hay or pasture seedings, new conservation structures, improvements (except as provided in Section V-B-5), or of applying lime and other long-lived fertilizers shall be divided between landlord and tenant as set forth in the following table. The tenant will be re-imbursed by the landlord either when the improvement is completed, **or** the tenant will be compensated for his share of the depreciated cost of his contribution when he leaves the farm based on the value of the tenant's contribution and depreciation rate shown in the following table. (Cross out the portion of the preceding sentence which does not apply.)

Rates for labor, power, and machinery contributed by the tenant shall be agreed upon before construction is started.

Compensation for Improvements Table

Type of improvement	Date to be completed	Estimated total cost (dollars)	Proportion to be contributed by tenant			Total value of tenant's contrib. (dollars) *	Rate of annual depreciation
			Material	Unskilled labor	Mach.		
			%	%	%		%

* To be recorded when improvement is completed.

2. Mineral rights. Nothing in this lease shall confer upon the tenant any right to minerals underlying said land, but same are hereby reserved by the landlord together with the full right to enter upon the premises and to bore, search, and excavate for same, to work and remove same, and to deposit excavated rubbish, and with full liberty to pass over said premises with vehicles and lay down and work any railroad track or tracks, tanks, pipelines, power lines, and structures as may be necessary or convenient for the above purpose. The landlord agrees to reimburse the tenant for any actual damage he may suffer for crops destroyed by these activities and to release the tenant from obligation to continue farming this property when development of mineral resources interferes materially with the tenant's opportunity to make a satisfactory return.

VII. FARM RECORDS AND FINANCIAL SETTLEMENTS

A. Records of joint interest shall be kept by ____________ and shall be made available to the landlord/tenant (cross out one) upon request. Financial and production records shall include a complete inventory of all property used in the farm business. Inventories shall be recorded and financial records summarized by ________ day of ______________ (month) or at intervals mutually agreed upon. Specify:

____

____

B. The record system to be used shall be:

____

____

C. All joint receipts and disbursements shall be handled through ____________________ bank as follows:

● Receipts: ______________________________

● Disbursements: _________________________

D. Cash financial settlement shall be made by the ______ day of each month or at intervals mutually-agreed upon. Specify:

____

____

VIII. ARBITRATION OF DIFFERENCES AND DIVISION OF PROPERTY

A. Arbitration of differences. Any differences between the parties as to their several rights or obligations under this lease that are not settled by mutual agreement after thorough discussion, shall be submitted for arbitration to a committee of three disinterested persons, one selected by each party hereto and the third by the two thus selected. The committee's decision shall be accepted by both parties.

B. Division of property. Upon termination of this lease, unused production shall be divided as follows:

1. Feed grain and supplies. All grain, silage, other feeds, and all co-owned supplies including straw and other bedding materials shall be divided by measure or value, whichever is more equitable, with the landlord and tenant each receiving title to his respective share as reported in Section V-A.

2. Livestock. If the livestock are owned equally (50-50), the tenant shall divide each class of livestock, as cows, steers, calves, hogs, etc., into two groups and the landlord shall take his choice of the two groups of each. In case the groupings cannot be made equal, a difference in monetary value shall be assigned before the choice is made and added to the choice.

3. Undivided interest of co-owned property. If both parties mutually agree not to accept the above described plan for dividing co-owned classes of property including livestock, it is agreed the tenant shall set a value on the entire amount of the respective co-owned classes of property on the basis of which he will either sell his undivided interest or buy that of the landlord, at the option of the landlord; or the co-owned property may be disposed of by private or public sale arranged for that purpose at a reasonable time and place.

4. Home use. The tenant and landlord may take annually for home use the following kinds and quantities of jointly owned crops, livestock, and/or livestock products:

____

____

____

Executed in duplicate on the date first above written:

_______________________________   _______________________________
              (tenant)   (landlord

_______________________________   _______________________________
          (tenant spouse)   (landlord spouse)

STATE OF _______________________________ }
                                          } SS:
COUNTY OF _______________________________ }

On this _______________________ day of _______________________________ A.D., 19______, before

me, the undersigned, a Notary Public in said State, personally appeared _______________________________

_______________________, _______________________, _______________________,

and _______________________, to me known to be the identical persons named in and who executed

the foregoing instrument, and acknowledged that they executed the same as their voluntary act and deed.

____
          (Notary Public)

Issued in furtherance of Cooperative Extension work, Acts of Congress of May 8 and June 30, 1914, in cooperation with the U.S. Department of Agriculture and Cooperative Extension Services of Illinois, Indiana, Kansas, Michigan, Minnesota, Missouri, Nebraska, North Dakota, Ohio, South Dakota and Wisconsin. John O. Dunbar, Director, Cooperative Extension Service, Kansas State University, Manhattan.

This PASTURE LEASE form can provide the landlord and tenant with a guide for developing an agreement to fit their individual situation. This form is not intended to take the place of legal advice pertaining to contractual relationships between the two parties. Because of the possibility that an operating agreement may be legally considered a partnership under certain conditions, seeking proper legal advice is recommended when developing such an agreement.

This lease is entered into this _____________ day of ________________________________, 19______, between

_________________________________, landlord, of ________________________________
(pasture owner)

____
(address)

_________________________________, spouse, of ________________________________

____
(address)

hereafter known as "the landlord," and

_________________________________, tenant, of ________________________________
(livestock owner)

____
(address)

_________________________________, spouse, of ________________________________

____
(address)

hereafter known as "the tenant."

I. PROPERTY DESCRIPTION

The landlord hereby leases to the tenant, to occupy and use for pasture purposes, the following described property:

____

____

____

____

____

consisting of approximately _____________ acres situated in _________________________ County (Counties), _________________________ (State) and on any other land which the landlord may designate by mutual written agreement.

II. GENERAL TERMS OF LEASE

A. Term.—[If a continuing lease is desired, use paragraph (1) and strike out (2).]

(1) Continuing Lease—The term of the lease shall be ________ year(s), commencing on the ___________ _____________ day of _____________________, 19______, and shall continue in effect from year to year thereafter (as an annual lease) unless written notice of termination is given by either party to the other at least _____________ days prior to expiration of this lease or the end of any year of continuation. If a definite term is desired, use paragraph (2) and strike out paragraph (1). No notice of termination is necessary if paragraph (2) is used.)

(2) Annual Lease—The term of this lease shall be _____________ year(s), commencing on the _____________ day of _____________________, 19______, and ending on the _____________ day of _____________________, 19______.

B. Review of lease.—A request for general review of the lease may be made, by either party at least ___________ days prior to the final date for giving notice to terminate the lease.

C. Amendments.—Amendments and alterations to this lease shall be in writing and shall be signed by both the landlord and tenant.

D. No partnerships created.—This lease shall not be deemed to give rise to a partnership relation, and neither party shall have authority to obligate the other without written consent, except as specifically provided in this lease.

E. Binding on Heirs.—The terms of this lease shall be binding upon the heirs, executors, administrators, and successors of both landlord and tenant in like manner as upon the original parties, except as provided by mutual written agreement otherwise.

F. Transfer of property.—-If the landlord should sell or otherwise transfer title to the farm, he will do so subject to the provisions of this lease.

G. Right of entry.—The landlord reserves the right of himself, his agents, his employees, or his assigns to enter the farm at any reasonable time for purposes (a) of consultation with the tenant; (b) of making repairs, improvements, and inspections; and (c) after notice of termination of the lease is given, of performing customary seasonal work, none of which is to interfere with the tenant in carrying out regular operations.

H. Additional agreements regarding term of lease:

____

____

____

____

____

I. Animal units (maximum allowable)—Not more than
_______________ animal units shall be kept in the
pasture at any one time without the express written
consent of the landlord. Deliberate violation of this
provision shall constitute grounds for termination of
this lease. (Each 1,000 pounds of average weight
shall be one animal unit. If the pasture owner and
the owner of the livestock prefer, they can use the
following basis for calculating animal units: 1 bull,
1.25 animal units; one 1,000-pound cow, 1 animal
unit; 1 yearling steer or heifer, .75 animal unit; calf,
6 months to 1 year, .5 animal unit; calf, 3 to 6
months, .3 animal unit; sheep, 5 per animal unit;
horse, 1.25 animal units.)—

Stocking Rate	Number Head	Number Animal Units
Bulls	_________	_________
Cows	_________	_________
Yearling steers	_________	_________
Yearling heifers	_________	_________
Calves, 6 mos.-1 year	_________	_________
Calves, 3-6 mos.	_________	_________
Other	_________	_________

III. OPERATION AND MAINTENANCE

A. The livestock owner agrees:

(1) Not to pasture livestock to be breachy. Should
any animal be found outside the pasture on at
least three occasions, the pasture owner may request its removal.

(2) Not to assign his right and duties under this
lease without the written consent of the pasture
owner.

(3) Not to put any cattle in pasture without getting
specific approval from the pasture owner in advance regarding number, health, sex, breed, and
age.

(4) Agrees to furnish health certificate as follows:

____
____
____
____
____
____

B. Both agree:

(1) Not to obligate other party. Neither party hereto
shall pledge the credit of the other party hereto
for any purpose whatsoever without the consent
of the other party. Neither party shall be responsible for debts or liabilities incurred, or for
damages caused by the other party.

(2) Responsibilities.—Additional responsibilities for
each party shall be divided as follows:

	Landlord	Tenant
Inspect fences not less than once per _________.	_____	_____
Furnish labor for repair of fences.	_____	_____
Furnish materials for repair of fences.	_____	_____
Supervise supply of water to livestock.	_____	_____
Furnish labor for repair of water system.	_____	_____
Materials for repair of water system.	_____	_____
Furnish salt and mineral.	_____	_____
Count livestock not less than once per _________.	_____	_____
Return stray animals to pasture.	_____	_____
Call veterinarian in case of emergency.	_____	_____
Pay veterinary expenses.	_____	_____
Provide loading and unloading facilities.	_____	_____
Furnish supplementary feed, if needed.	_____	_____
Notify other party of shortage in count.	_____	_____
Provide facilities for fly control.	_____	_____
Keep fly control facilities in working order.	_____	_____
Liability Insurance.	_____	_____

(3) Additional agreements:

____
____
____
____
____

IV. RENTAL CALCULATIONS AND PAYMENT SCHEDULE

(Use Method I, II or III and Strike Out the Two Methods Not Used)

METHOD I—The tenant owner agrees to pay $________
per acre for use of the property described in paragraph I.
Total rent of $____________ shall be paid as follows.

$________ on or before ______ day of ________ (month),
$________ on or before ______ day of ________ (month),
$________ on or before ______ day of ________ (month),
$________ on or before ______ day of ________ (month),

If rent is not paid when due, the tenant agrees to pay
interest on the amount of unpaid rent at the rate of
________ percent per annum from the due date until paid.

Rental adjustment—Additional agreements in regard
to rental payment:

____
____
____
____
____

METHOD II—The livestock owner agrees to pay the following rates: (The period may be by the month, pasture season or year.)

	Number	X	Rental Rate/Period	=	Total Rent/Period
Bulls	_____________	X	$____________	=	$____________
Cows	_____________	X	$____________	=	$____________
Yearling steers	_____________	X	$____________	=	$____________
Yearling heifers	_____________	X	$____________	=	$____________
Calves, 6 mos.-1 year	_____________	X	$____________	=	$____________
Calves, 3-6 mos.	_____________	X	$____________	=	$____________
Other	_____________	X	$____________	=	$____________
Total Rent					$____________

The minimum rent shall be $_______________. Such rental shall be required regardless of whether or not livestock are actually being pastured. The Total Rent of $____________ shall be paid as follows.

$_______ on or before _______ day of _______ (month),

$_______ on or before _______ day of _______ (month),

$_______ on or before _______ day of _______ (month),

$_______ on or before _______ day of _______ (month),

If rent is not paid when due, the tenant agrees to pay interest on the amount of unpaid rent at the rate of _______ percent per annum from the due date until paid.

Rental adjustment—Additional agreements in regard to rental payment:

____

____

____

____

____

____

____

____

____

____

METHOD III—Other Rental Arrangements (Share of gain-etc.)

____

____

____

____

____

____

____

VI. ARBITRATION OF DIFFERENCES

Any differences between the parties as to their several rights or obligations under this lease that are not settled by mutual agreement after thorough discussion, shall be submitted for arbitration to a committee of three disinterested persons, one selected by each party hereto and the third by the two thus selected. The committee's decision shall be accepted by both parties.

Executed in duplicate on the date first above written:

_______________________________   _______________________________
tenant (Livestock owner)   landlord (Pasture owner)

_______________________________   _______________________________
(tenant spouse)   (landlord spouse)

COUNTY OF _______________________ ⎫
                                 ⎬ SS:
STATE OF _______________________ ⎭

On this _______________________ day of _______________________ A.D., 19_____, before
me, the undersigned, a Notary Public in said State, personally appeared _______________________
_______________________, _______________________, _______________________,
and _______________________, to me known to be the identical persons named in and who executed
the foregoing instrument, and acknowledged that they executed the same as their voluntary act and deed.

____
Notary Public

Issued in furtherance of Cooperative Extension work, Acts of Congress of May 8 and June 30, 1914, in cooperation with the U.S. Department of Agriculture and Cooperative Extension Services of Illinois, Indiana, Kansas, Michigan, Minnesota, Missouri, Nebraska, North Dakota, Ohio, South Dakota and Wisconsin. John O. Dunbar, Director, Cooperative Extension Service, Kansas State University, Manhattan.

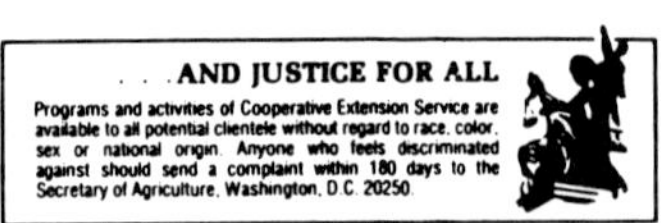

Lease Supplement for Making Improvements on a Rented Farm

(Prepared by E. G. Stoneberg, extension economist, Iowa State University)

Description of Farm: Section__________ Township__________ Range__________ County__________ Size__________ Acres more or less__________

1. In consideration of the agreements herein contained, the signers agree that the improvements listed in Section A (below) will be completed on the above-described farm on or before the date listed in Section B.

2. It is agreed that the signers will share contributions and costs necessary to the completion of these improvements as set forth in Section C.

3. It is agreed that the estimated value or cost of the tenant's contributions will be listed in Section D.

4. It is further agreed that the estimated value or cost of the tenant's contributions will be depreciated at the uniform annual percentage rate listed in Section E. The year of first depreciation is to be listed in Section F.

5. If for any reason the tenant leaves the farm before the tenant's estimated value or cost (Sec. D) is fully recovered through annual use and depreciation (Sec. E), then the landlord will pay the tenant for the remaining undepreciated value of the tenant's investment.

6. It is agreed that each item as set forth opposite the signatures of the landlord and tenant will be viewed as a separate contract supplemental to the lease. New items may be agreed upon at any time during the term of the lease and recorded in the spaces below.

Section A Type and location of improvement	Section B Date to be completed	Section C Percentage of contributions assumed by landlord (L) by tenant (T)						Section D Estimated value or cost of tenant's investment	Section E Annual rate of depreciation (percent)	Section F Lease year when depreciation begins	Section G Date signed	Section H—Signatures I hereby accept my indicated share of the responsibility for the improvements recorded in Section A, which I have approved.
		Materials		Labor		Machinery or trucking						
		L	T	L	T	L	T					
												L.
												T.
												L.
												T.
												L.
												T.
												L.
												T.
												L.
												T.
												L.
												T.
												L.
												T.
												L.
												T.